Technology for the United States Navy and Marine Corps, 2000-2035

Becoming a 21st-Century Force

VOLUME 1 Overview

Committee on Technology for Future Naval Forces
Naval Studies Board
Commission on Physical Sciences, Mathematics, and Applications
National Research Council

NATIONAL ACADEMY PRESS
Washington, D.C.   1997

NOTICE: The project that is the subject of this report was approved by the Governing Board of the National Research Council, whose members are drawn from the councils of the National Academy of Sciences, the National Academy of Engineering, and the Institute of Medicine. The members of the committee responsible for the report were chosen for their special competences and with regard for appropriate balance.

This report has been reviewed by a group other than the authors according to procedures approved by a Report Review Committee consisting of members of the National Academy of Sciences, the National Academy of Engineering, and the Institute of Medicine.

The National Academy of Sciences is a private, nonprofit, self-perpetuating society of distinguished scholars engaged in scientific and engineering research, dedicated to the furtherance of science and technology and to their use for the general welfare. Upon the authority of the charter granted to it by the Congress in 1863, the Academy has a mandate that requires it to advise the federal government on scientific and technical matters. Dr. Bruce Alberts is president of the National Academy of Sciences.

The National Academy of Engineering was established in 1964, under the charter of the National Academy of Sciences, as a parallel organization of outstanding engineers. It is autonomous in its administration and in the selection of its members, sharing with the National Academy of Sciences the responsibility for advising the federal government. The National Academy of Engineering also sponsors engineering programs aimed at meeting national needs, encourages education and research, and recognizes the superior achievements of engineers. Dr. William A. Wulf is president of the National Academy of Engineering.

The Institute of Medicine was established in 1970 by the National Academy of Sciences to secure the services of eminent members of appropriate professions in the examination of policy matters pertaining to the health of the public. The Institute acts under the responsibility given to the National Academy of Sciences by its congressional charter to be an adviser to the federal government and, upon its own initiative, to identify issues of medical care, research, and education. Dr. Kenneth I. Shine is president of the Institute of Medicine.

The National Research Council was organized by the National Academy of Sciences in 1916 to associate the broad community of science and technology with the Academy's purposes of furthering knowledge and advising the federal government. Functioning in accordance with general policies determined by the Academy, the Council has become the principal operating agency of both the National Academy of Sciences and the National Academy of Engineering in providing services to the government, the public, and the scientific and engineering communities. The Council is administered jointly by both Academies and the Institute of Medicine. Dr. Bruce Alberts and Dr. William A. Wulf are chairman and vice chairman, respectively, of the National Research Council.

This work was performed under Department of the Navy Contract N00014-96-D-0169/0001 issued by the Office of Naval Research under contract authority NR 201-124. However, the content does not necessarily reflect the position or the policy of the Department of the Navy or the government, and no official endorsement should be inferred.

The United States Government has at least a royalty-free, nonexclusive, and irrevocable license throughout the world for government purposes to publish, translate, reproduce, deliver, perform, and dispose of all or any of this work, and to authorize others so to do.

Copyright 1997 by the National Academy of Sciences. All rights reserved.

Copies available from:

Naval Studies Board
National Research Council
2101 Constitution Avenue, N.W.
Washington, D.C. 20418

Printed in the United States of America

COMMITTEE ON TECHNOLOGY FOR FUTURE NAVAL FORCES

DAVID R. HEEBNER, Science Applications International Corporation (retired), *Study Director*
ALBERT J. BACIOCCO, JR., The Baciocco Group, Inc.
ALAN BERMAN, Applied Research Laboratory, Pennsylvania State University
NORMAN E. BETAQUE, Logistics Management Institute
GERALD A. CANN, Raytheon Company
GEORGE F. CARRIER, Harvard University
SEYMOUR J. DEITCHMAN, Institute for Defense Analyses (retired)
ALEXANDER FLAX, Potomac, Maryland
WILLIAM J. MORAN, Redwood City, California
ROBERT J. MURRAY, Center for Naval Analyses
ROBERT B. OAKLEY, National Defense University
JOSEPH B. REAGAN, Saratoga, California
VINCENT VITTO, Lincoln Laboratory, Massachusetts Institute of Technology

Navy Liaison Representatives
RADM JOHN W. CRAINE, JR., USN, Office of the Chief of Naval Operations, N81 (as of July 4, 1996)
VADM THOMAS B. FARGO, USN, Office of the Chief of Naval Operations, N81 (through July 3, 1996)
RADM RICHARD A. RIDDELL, USN, Office of the Chief of Naval Operations, N91
CDR DOUGLASS BIESEL, USN, Office of the Chief of Naval Operations, N812C1

Marine Corps Liaison Representative
LtGen PAUL K. VAN RIPER, USMC, Marine Corps Combat Development Command

Consultants
LEE M. HUNT
SIDNEY G. REED, JR.
JAMES G. WILSON

Staff
RONALD D. TAYLOR, Director, Naval Studies Board
PETER W. ROONEY, Program Officer
SUSAN G. CAMPBELL, Administrative Assistant
MARY G. GORDON, Information Officer
CHRISTOPHER A. HANNA, Project Assistant

NAVAL STUDIES BOARD

DAVID R. HEEBNER, Science Applications International Corporation (retired), *Chair*
GEORGE M. WHITESIDES, Harvard University, *Vice Chair*
ALBERT J. BACIOCCO, JR., The Baciocco Group, Inc.
ALAN BERMAN, Applied Research Laboratory, Pennsylvania State University
NORMAN E. BETAQUE, Logistics Management Institute
NORVAL L. BROOME, Mitre Corporation
GERALD A. CANN, Raytheon Company
SEYMOUR J. DEITCHMAN, Institute for Defense Analyses (retired), *Special Advisor*
ANTHONY J. DeMARIA, DeMaria ElectroOptics Systems, Inc.
JOHN F. EGAN, Lockheed Martin Corporation
ROBERT HUMMEL, Courant Institute of Mathematical Sciences, New York University
DAVID W. McCALL, Far Hills, New Jersey
ROBERT J. MURRAY, Center for Naval Analyses
ROBERT B. OAKLEY, National Defense University
WILLIAM J. PHILLIPS, Northstar Associates, Inc.
MARA G. PRENTISS, Jefferson Laboratory, Harvard University
HERBERT RABIN, University of Maryland
JULIE JCH RYAN, Booz, Allen and Hamilton
HARRISON SHULL, Monterey, California
KEITH A. SMITH, Vienna, Virginia
ROBERT C. SPINDEL, Applied Physics Laboratory, University of Washington
DAVID L. STANFORD, Science Applications International Corporation
H. GREGORY TORNATORE, Applied Physics Laboratory, Johns Hopkins University
J. PACE VanDEVENDER, Prosperity Institute
VINCENT VITTO, Lincoln Laboratory, Massachusetts Institute of Technology
BRUCE WALD, Arlington Education Consultants

Navy Liaison Representatives
RADM JOHN W. CRAINE, JR., USN, Office of the Chief of Naval Operations, N81 (as of July 4, 1996)
VADM THOMAS B. FARGO, USN, Office of the Chief of Naval Operations, N81 (through July 3, 1996)
RADM RICHARD A. RIDDELL, USN, Office of the Chief of Naval Operations, N91
RONALD N. KOSTOFF, Office of Naval Research

Marine Corps Liaison Representative
LtGen PAUL K. VAN RIPER, USMC, Marine Corps Combat Development Command

RONALD D. TAYLOR, Director
PETER W. ROONEY, Program Officer
SUSAN G. CAMPBELL, Administrative Assistant
MARY G. GORDON, Information Officer
CHRISTOPHER A. HANNA, Project Assistant

COMMISSION ON PHYSICAL SCIENCES, MATHEMATICS, AND APPLICATIONS

ROBERT J. HERMANN, United Technologies Corporation, *Co-Chair*
W. CARL LINEBERGER, University of Colorado, *Co-Chair*
PETER M. BANKS, Environmental Research Institute of Michigan
LAWRENCE D. BROWN, University of Pennsylvania
RONALD G. DOUGLAS, Texas A&M University
JOHN E. ESTES, University of California at Santa Barbara
L. LOUIS HEGEDUS, Elf Atochem North America, Inc.
JOHN E. HOPCROFT, Cornell University
RHONDA J. HUGHES, Bryn Mawr College
SHIRLEY A. JACKSON, U.S. Nuclear Regulatory Commission
KENNETH H. KELLER, University of Minnesota
KENNETH I. KELLERMANN, National Radio Astronomy Observatory
MARGARET G. KIVELSON, University of California at Los Angeles
DANIEL KLEPPNER, Massachusetts Institute of Technology
JOHN KREICK, Sanders, a Lockheed Martin Company
MARSHA I. LESTER, University of Pennsylvania
THOMAS A. PRINCE, California Institute of Technology
NICHOLAS P. SAMIOS, Brookhaven National Laboratory
L.E. SCRIVEN, University of Minnesota
SHMUEL WINOGRAD, IBM T.J. Watson Research Center
CHARLES A. ZRAKET, Mitre Corporation (retired)

NORMAN METZGER, Executive Director

Preface

This study was inspired by the events of the past decade, which saw a vast transformation in the international strategic landscape facing the United States, and in the missions and perspectives of the U.S. Navy and Marine Corps as implementing arms of U.S. national security strategy. The terms of reference of the study, developed by VADM Thomas B. Fargo, USN, and RADM Richard A. Riddell, USN, and signed by the Chief of Naval Operations on November 28, 1995, requested that the National Research Council undertake a thorough examination of the impact of advancing technology on the form and capability of the naval forces to the year 2035. Recognizing the anticipated austere budget environment, the terms of reference sought leverage to increase the cost-effectiveness of those forces in that environment, in many technical areas. They specifically asked for an identification of "present and emerging technologies that relate to the full breadth of Navy and Marine Corps mission capabilities," with specific attention to "(1) information warfare, electronic warfare, and the use of surveillance assets; (2) mine warfare and submarine warfare; (3) Navy and Marine Corps weaponry in the context of effectiveness on target; [and] (4) issues in caring for and maximizing effectiveness of Navy and Marine Corps human resources." Ten specific technical areas were identified to which attention should be broadly directed. The terms of reference are given in full in Appendix A of this report.

These terms of reference follow from a 1988 study of similar scope, the Navy-21 study,[1] that covered much the same ground, but in the earlier context of the Cold War that was still ongoing. At the completion of the Navy-21 study, it

---

[1] Naval Studies Board. 1988. *Navy-21: Implications of Advancing Technology for Naval Operations in the Twenty-First Century,* National Academy Press, Washington, D.C.

was recognized that the results of a study as broad as that would have to be reviewed periodically to see what had changed in the international security environment, in pertinent domestic circumstances, and in technology, and to renew the projections in the light of those changes. An earlier update of the Navy-21 study,[2] and additional Naval Studies Board studies in the areas of advanced sensing,[3] mine warfare,[4] the combat information network,[5] future aircraft carriers,[6] command, control, and communications for strike warfare,[7] shipboard waste disposal,[8] the Navy and Marine Corps in regional conflict,[9] and conflict deterrence in the post-Cold War world,[10] bore on topics related to naval force development since the publication of the Navy-21 study. All of these studies contributed to the background of the current study, and indeed their results informed the current study in many areas.

To carry out this study, eight technical panels were organized to examine all of the specific technical areas called out in the terms of reference, with some of the 10 topics combined under the cognizance of individual panels as the logic of the topics suggested. The panel structure of the study is shown in Figure P.1. Altogether, some 130 experts in the various technical areas participated in the study as panel members, senior advisors, or participants invited to help the panels with specific tasks. In addition, about 30 Navy and Marine Corps liaison representatives met frequently with the technical panels and with the total study membership during the course of the study. They contributed essential support in providing necessary information and in helping the panel members and leadership understand ongoing Service programs and policies. All of the study participants and Service representatives are listed, with the panels they contributed to, in Appendix B.

---

[2] Naval Studies Board. 1993. *Navy-21 Update: Implications of Advancing Technology for Naval Operations in the Twenty-First Century,* National Academy Press, Washington, D.C.

[3] Naval Studies Board. 1985. *Sensor Panel Report, Phase II (U),* National Academy Press, Washington, D.C. (Classified).

[4] Naval Studies Board. 1992-1993. *Mine Countermeasures Technology,* Vol. I-IV, National Academy Press, Washington, D.C.

[5] Naval Studies Board. 1991. *Combat Networks for Distributed Naval Forces (U),* National Academy Press, Washington, D.C. (Classified).

[6] Naval Studies Board. 1991. *Carrier-21: Future Aircraft Carrier Technology,* National Academy Press, Washington, D.C.

[7] Naval Studies Board. 1994. *Naval Communications Architecture,* National Academy Press, Washington, D.C.

[8] Naval Studies Board. 1996. *Shipboard Pollution Control: U.S. Navy Compliance With MARPOL Annex V,* National Academy Press, Washington, D.C.

[9] Naval Studies Board. 1996. *The Navy and Marine Corps in Regional Conflict in the 21st Century,* National Academy Press, Washington, D.C.

[10] Naval Studies Board. 1997. *Post-Cold War Conflict Deterrence,* National Academy Press, Washington, D.C.

PREFACE

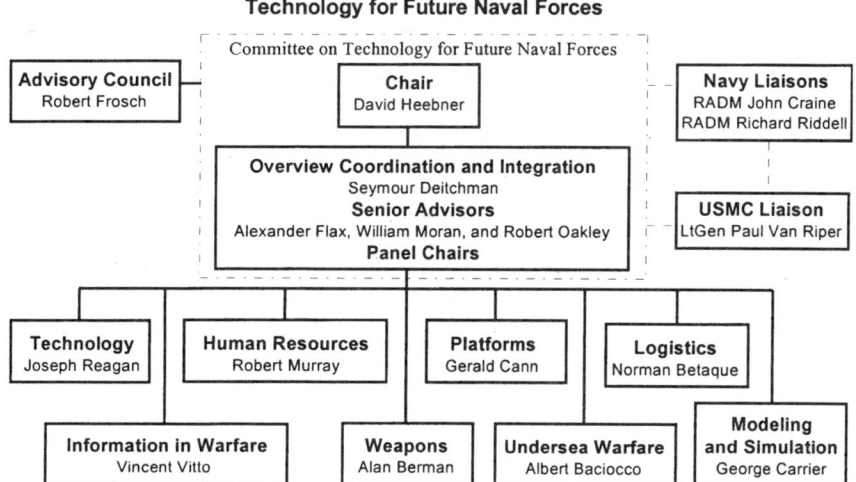

FIGURE P.1 Organizational structure.

Some 100 panel meetings were held during the course of the study, during which there were briefings by the Services and industry, and working sessions to arrive at the study results. Projections of the international security environment, the relationships of the diverse panel outputs to each other, and the significance of those outputs for the naval forces were brought together and interpreted by a coordination and integration group composed of a chairman, the three senior advisors to the study, and the chairmen of the eight technical panels. This group was constituted as the Committee on Technology for Future Naval Forces. The members of the committee met bimonthly to inform each other of progress in the individual panels' efforts and to resolve issues that cut across the responsibilities of more than one panel as they emerged during the panels' work. This overview report is the result of the committee's efforts.

There were three plenary sessions of the entire study membership. The first, in March 1996, was addressed by the Chief of Naval Operations and many high-level officials of the Navy Department, the other Services, the Defense Department, and industry. This served as an organization meeting and conveyed a common, starting information base to the study membership. At the second plenary session, in October 1996, all the members of the study had their first opportunity to review each other's work, to see how the results of all the panels' work were coming together into an integrated overview, and to feed the results back into their own efforts. The last plenary session, in March 1997, served as a coordination and writing session in which all of the panels' reports and this overview report were completed for final review and checked to ensure that the overview and the eight panel reports were consistent with each other and mutually supporting. This overview, the first volume in the nine-volume series produced as a result of this study, is, of course, based on the detailed material developed by the individual eight panels. That material is presented in eight separately published volumes whose major topics are outlined in Box P.1.

A periodic "sanity check" on the progress and results of the entire study was provided by the Advisory Council, which met three times during the course of the study: early in the study, to review and advise on the study plans and scope as they were developing within the panels; after the first plenary session, to review and advise on the panels' outputs and on the integrated results as they began to appear; and after the last plenary session, to advise on the final results of the study as they emerged from that session. This final version of the overview report and the final versions of the panel reports reflect the Advisory Council's inputs as well as the comments made in the National Research Council review process.

Throughout this report, the term "naval forces" is used to refer to the Navy and the Marine Corps together. Much of the discussion in the report refers to the two Services. Especially in the post-Cold War world, the two Services are inextricably linked together and must function essentially as a single force over much of their mission spectrum. This is especially true in the difficult and complex transfer of Marine combat power from the sea to the shore against actual or potential opposition, and in the subsequent support of forces ashore by sea-based firepower and logistics. Although several of the important system advances described in this report will clearly apply more to the Navy than to the Marine Corps, the developing Marine Corps concept for Operational Maneuver From the Sea (OMFTS), the associated Marine Corps systems, and the required Navy firepower and logistic support were examined in some detail in the recent Naval Studies Board report on regional conflict.[11] That report examines many system and technology issues that are germane to the subjects discussed in this report. The two reports should be considered companion pieces that, together, probe the future system and technology needs of the two Services that make up the naval forces.

Many more areas of effort and investment than could be treated in this overview report are reviewed and presented in the reports of the study panels, in accordance with their specific areas of concern. Omission of an item from this overview does not imply a judgment on its relative or absolute importance. In many cases, the level of detail precluded detailed treatment here and resulted in the item's being subsumed in a more general topic area. Also, this first volume in the nine-part series concentrates on the presentation of new or different ideas for technology and systems, and their impact on naval force operations; programs

---

[11] Naval Studies Board. 1996. *The Navy and Marine Corps in Regional Conflict in the 21st Century*, National Academy Press, Washington, D.C.

under way that need no comment in that respect are noted in context where appropriate and essential to the discussion, but they are not elaborated in this report.

The subject of the use of nuclear weapons by the naval forces was raised during the conduct of this study in connection with the need for destruction of deeply buried targets that conventional weapons may not be able to reach. Such targets might contain key command centers, or they might be storage sites for weapons of mass destruction. Deeply buried command centers can be inactivated for practical use by conventional attacks against exposed support systems (for ventilation, power, communications, and so on). The key problem would be gaining intelligence on these systems to guide the attack. Also, access to storage sites might be cut off by similar means. However, physical destruction of stored agents or warheads in deeply buried sites may require nuclear weapons. In addition to the possible use of existing weapons for such purposes, the very high accuracies that will be achievable in the future may make it possible to attack such targets with much smaller nuclear warheads than are currently available. Nuclear weapons will be available to the military forces for many years to come, even in the presence of arms control agreements extant and to be negotiated in the future, but their use will be governed by national security decisions at the highest levels of government. In case it is desired to design and build a new class of much smaller nuclear warheads than those now available, they could not be tested under the Comprehensive Test Ban Treaty. These issues are extraordinarily complex, with political dimensions far exceeding their technical aspects. They affect all of the armed forces and national policies in many national security-related areas. They must be dealt with at the national level, and they continue to require attention. Although they arose and were discussed during this study, no attempt was made in this report to deal in a substantive way with the broad issues involving nuclear weapons. Some of those issues are discussed in the recent Naval Studies Board study on post-Cold War conflict deterrence.[12]

This overview report is unclassified, to permit the greatest possible circulation to the communities concerned with the future of the naval forces and to the interested public. However, the study participants have taken pains to ensure that, to the best of their knowledge, the unclassified results of this study remain consistent with classified research and development efforts being conducted by the Department of the Navy and other agencies that will contribute to future naval force capabilities in important ways.

The results of this nine-volume study show, as requested in the terms of reference, research and development paths by which the naval forces can become as capable and responsive as will be expected of them in the future, within anticipated fiscal constraints. By itself, however, the ability of the forces to respond

---

[12] Naval Studies Board. 1997. *Post-Cold War Conflict Deterrence,* National Academy Press, Washington, D.C.

> **BOX P.1**
> **Topics Addressed in Volumes 2 Through 9**
>
> **Volume 2: Technology**
> - Discusses nine technology areas expected to be of greatest importance for future naval operations, including computation, information and communications, sensors, automation, human performance, materials, power and propulsion, environmental technology, and enterprise processes
> - Emphasizes the continuing key role for Navy-sponsored R&D in sensors, special materials, fluid dynamics, ship power and propulsion, and oceanography
>
> **Volume 3: Information in Warfare**
> - Discusses offensive and defensive information warfare
> - Emphasizes the dependence of future naval operations and capability on commercial information technology and infrastructure
> - Discusses the role of advanced sensors in information collection; points out the importance of naval space-based operations
> - Defines and discusses aspects of strategy for achieving information superiority in warfare
>
> **Volume 4: Human Resources**
> - Presents strategic objectives for developing and maintaining human performance and competence in naval operations
> - Discusses the importance of information-based training and job performance enhancement
> - Points out recruitment opportunity in 2-year colleges
> - Outlines quality-of-life issues and the need to assess results of efforts toward improvement
>
> **Volume 5: Weapons**
> - Covers the following categories: offensive and defensive systems, surface-to-surface and air-to-surface weapons, air-to-air weapons, weapons for undersea warfare, laser weapons, special-purpose weapons, and sea-based nuclear weapon alternatives
> - Discusses the following concepts: family of low-cost, modular, rocket-propelled, precision-guided, sea-based missiles for land attack; explosive and propellant improvements to decrease the size of all munitions, missiles, and torpedoes; networked multimode targeting for cooperative antimissile defense; networked sensors for distributed minefields for offensive and defensive mine warfare; long-range missiles for air-to-air combat; theater ballistic missile defense needed for amphibious operations; laser weapons for aircraft defense; and special systems for urban combat and hard targets
>
> **Volume 6: Platforms**
> - Emphasizes pursuit of R&D in common technology thrust areas: stealth, automation, minimal manning, affordability, fluid and flow control, and off-board unmanned vehicles; acknowledges environmental issues

- Discusses the following:
  - Surface ships: automation and integrated information and control systems, passive and active signature management, and modular electric propulsion
  - Aircraft: increasingly short takeoff and landing (STOL) and short takeoff and vertical landing (STOVL), Integrated High Performance Turbine Engine Technology (IHPTET) program engine improvements, improved infrared stealth, and cost control with advanced design and manufacturing
  - Submarines: continued stealth, integrated payload systems for power projection, improved sensors and connectivity for cooperative engagement capability and strike, and higher-power-density propulsion

Volume 7: Undersea Warfare
- Antisubmarine warfare (ASW)
  - Points out increasing ASW threat; at-sea training exercises essential
  - Emphasizes importance of advances in computation, sensors, and oceanography to regain much of U.S. acoustic advantage
  - Discusses use of multiple platforms, a cooperative engagement capability-like network, for littoral ASW
- Mine warfare and mine countermeasures (MCM): discusses more rapid MCM with a balanced system involving integrated ISR; organic capability with small transportable SWATH ship; night helicopter MCM operations and expendable neutralizers; shallow water surveillance and networking with UUVs, mammals, EODs, and SEAL teams; and brute-force surf and beach breaching using controlled space-time explosive patterns

Volume 8: Logistics
- Discusses use of information-based systems for total asset visibility and control for managing and moving materiel
- Discusses prospective use of information technology for reducing maintenance and for supporting weapon system readiness
- Points out the advantages of using modeling and simulation in designing a fully functional logistic ship
- For logistics hardware, emphasizes more containerization, robotics for rough-sea cargo transfer, methods for VLS reload, and improved vehicles for ship-to-shore transport

Volume 9: Modeling and Simulation
- Describes current modeling and simulation (M&S) infrastructure
- Surveys prospects for M&S technology developments, including those for decision support, acquisition, and training
- Describes potential pitfalls in naval use of M&S, especially those related to model validity and system complexity
- Addresses challenges in assimilating and exploiting M&S technology
- Presents an approach to prioritizing warfare subjects for research
- Describes needed improvements in the conceptual, methodological, and technological infrastructure for M&S

rapidly and effectively to crises requiring military action will not be a sufficient condition for crisis resolution. The government decision processes leading up to commitment of the forces to a mission could well be a pacing detail that could dissipate the advantages of the naval forces' strength and responsiveness for crisis resolution.

Finally, the Naval Studies Board wishes to express its appreciation to the dozens of Navy, Marine Corps, other Service, Defense Department, and commercial industry representatives who contributed information to this study and who briefed the panels during their meetings. Without these inputs the essential information and perspectives on the future national security environment and developing future technology could not have been developed.

Contents

PART I: SYNOPSIS

1 THE 2035 NAVAL FORCES 3
   Naval Force Missions, 3
   Driving Factors in the Future Naval Force Environment, 4
   The Future Naval Forces, 5

2 CREATING THE 2035 NAVAL FORCES 8
   Naval Force Restructuring, 8
   Future Naval Force Capabilities, 10
   Research and Development, 19

3 RECOMMENDATIONS 20

PART II: OVERVIEW DISCUSSION

4 INTRODUCTION 25

5 THE INTERNATIONAL SECURITY ENVIRONMENT: 2000–2035 ... 30
   International Security Trends, 30
   Additional Factors in the Environment, 34
   Military Capabilities of Potential Adversaries, 36
   Strategic Significance of the Future Naval Force Environment, 39

6 ANTICIPATED U.S. NAVAL FORCE CAPABILITIES: 2000–2035 42
   The Emerging Shape of the Future Naval Forces, 42
   What Will the Naval Forces Be Required to Do?, 48
   How Will the Naval Forces Operate?, 50

7 ENTERING WEDGES OF CAPABILITY TO SHAPE THE
      NAVAL FORCES OF 2000 TO 2035 52
   Information Systems and Operations, 54
   Enhancing the Capabilities of Individual Sailors and Marines, 60
   The Combat Fleet, 64
   New Approaches to Undersea Warfare, 77
   New Approaches to Operations in Populated Areas, 84
   Reengineering the Logistic System, 87
   Modeling and Simulation as a Foundation Technology, 92
   Focused Research and Development, 95

8 IMPLICATIONS FOR THE DEPARTMENT OF THE NAVY 101
   A Conceptual Revolution, 101
   Payoffs and Vulnerabilities, 102
   Implications for Naval Force Planning, 105
   An Evolutionary Approach to Revolutionary Capability, 106

APPENDIXES

A Terms of Reference 111

B Study Membership and Participants 116

C Acronyms and Abbreviations 124

Part I
Synopsis

1

The 2035 Naval Forces

The future national security environment in which the naval forces will play a key part is likely to change much more rapidly than the naval forces themselves can be changed. A great deal of adaptability must therefore be incorporated into them from the start. Their form, modes of operation, and military capability will also be driven in large part by the rapidly advancing technology that will build them. This study explores the nature of the future environment in which U.S. naval forces will have to operate, and it examines how technology can be applied to restructuring the naval forces to better position them to meet the challenges of that environment.

NAVAL FORCE MISSIONS

The tasks that naval forces are required to perform have changed little over the decades and are expected to continue in the future. They will include:

- Sustaining a forward presence;
- Establishing and maintaining blockades;
- Deterring and defeating attacks on the United States, our allies, and friendly nations, and, in particular, sustaining a sea-based nuclear deterrent force;
- Projecting national military power through modern expeditionary warfare, including attacking land targets from the sea, landing forces ashore and providing fire and logistic support for them, and engaging in sustained combat when necessary;
- Ensuring global freedom of the seas, airspace, and space; and
- Operating in joint and combined settings in all these missions.

- Nuclear, chemical, and biological weapons.

The naval forces will have to be designed to meet the kinds of opposition these threats will pose.

THE FUTURE NAVAL FORCES

Future Naval Force Capability

Application of the advancing technologies that are described in Chapter 6 of this report can lead to a complete transformation of the naval forces, amounting to a breakthrough in naval force capabilities. Foremost among the breakthrough capabilities that could be achieved are the following:

- Sustained information superiority over adversaries;
- Major ships operated effectively by many fewer people, through the use of networked instrumentation and automated subsystems;
- A family of rocket-propelled, guided missiles, significantly lower in cost than today's weapons, that will greatly increase the responsiveness, rate of fire, volume of fire, and accuracy of strike, interdiction, and supporting fire from surface combatants and submarines;
- STOL or STOVL, stealth, and standoff in combat aircraft;
- Cooperative air-to-air engagement at long range using networked multistatic sensor, aircraft, and missile systems;
- Use of unmanned aerial vehicles (UAVs) for both routine and excessively dangerous tasks;
- Greatly expanded submarine capability to support naval force operations ashore;
- Recapture of the antisubmarine warfare advantage that has been eroded by quieting of Russian nuclear submarines and by advanced air-independent nonnuclear submarines that are being sold by other nations on world markets;
- The ability to negate minefields at sea, in the surf, and on the beaches much more rapidly than has been possible heretofore;
- Novel weapons, systems, and techniques for fighting in populated areas, against organized military forces, irregulars, and terrorist and criminal groups; and
- Logistic support extensively based at sea that will provide needed materiel on time with far less excess supply in the system than has been the case in the past.

Operations during the Gulf War and since have shown that such capabilities, in the main, are still in rudimentary form. Broad implementation of all of them in integrated fashion must await full development, maturation, and application of the technologies described. This series of changes will add up to a revolution over the 40 years envisioned by the study.

Future Naval Force Operations

Naval force operations using these new capabilities will be characterized by the following:

- Operations from forward deployment, with a few major, secure bases of prepositioned equipment and supplies;
- Great economy of force based on early, reliable intelligence; on the timely acquisition, processing, and dissemination of local, conflict-, and environment-related information; and on all aspects of information warfare;
- Combined arms operations from dispersed positions, using stealth, surprise, speed, and precision in identifying targets and attacking opponents, with fire and forces massed rapidly from great distances at decisive locations and times;
- Defensive combat operations and systems, from ship self-defense through air defense, antisubmarine warfare, and antitactical ballistic missile defense, always networked in cooperative engagement modes that extend from the fleet to cover troops and installations ashore;
- Marine Corps operations in dispersed, highly mobile units from farther out at sea to deeper inland over a broader front, with more rapid conquest or neutralization of hostile populated areas, in the mode currently evolving into the doctrine for Operational Maneuver From the Sea;
- Extensive use of commercial firms for maintenance and support functions; and
- Extensive task sharing and mission integration in the joint and combined environment, with many key systems, especially in the information area, jointly operated. (Operational "fallback" positions for naval forces whose joint support is delayed or prevented from arriving by the exigencies of conflict are discussed in Chapter 8, in the section titled "Payoffs and Vulnerabilities.")

Synthesis

Taken together with ongoing work on defense against cruise missiles and tactical or theater ballistic missile defenses, the vision evoked by these advanced capabilities, if they are implemented and used to enable leaner, more streamlined modes of operation, can position the naval forces to carry out their missions in the face of future international security challenges, threats, and fiscal constraints far more efficiently and effectively than today's forces could. Implementing the capabilities will require a stable, sustained R&D program, in areas that are described later in this report. Modeling and simulation (M&S) has become a foundation technology in naval system and force development and utilization. If developed in directions described in this report and used appropriately, this technology can greatly facilitate progress toward the goals described above; indeed, in some areas the capability sought will be difficult or impossible to achieve within reasonable resource expenditures without the use of modeling and simulation.

The force development described will have to proceed on many fronts simultaneously. Otherwise, delays in advancing some capabilities—such as failure to establish information superiority, or to develop the responsive firepower needed to support dispersed forces ashore, or to meet the threats of mines, submarines, and missiles, or to be able to dominate populated areas quickly, or to advance the logistic system together with the combat systems—can turn into "showstoppers" for the entire naval force.

The resulting "lean" forces will inevitably have vulnerabilities that must be accounted for. The most serious of these will emerge from disruption of operations due to enemy action and the well-known "fog" and "friction" of war, and from failure of key force elements to perform when expected and as expected, for unforeseen reasons. Prudent steps (detailed in Chapter 8 of this report) can be taken to mitigate the worst effects of the vulnerabilities. Such mitigation efforts must be built into the system and force design. The character and cost of such "insurance" programs must be considered an integral part of the effort in implementing the new naval force capabilities.

2

Creating the 2035 Naval Forces

NAVAL FORCE RESTRUCTURING

To meet the demands of the future environment with the capabilities technology will make available, the naval forces will have to be extensively restructured—not instantly, but over time. The results of this study suggest that the following steps will be necessary:

1. The Navy and the Marine Corps must make joint modernization plans based on jointly formulated concepts of operation; their missions are overlapping and complementary, and they will be operating and fighting together much of the time.

2. The new kinds of forces can be created by investing in "entering wedges" of capability (the main ones are outlined below) that the forces can work with and learn how to use. To manage technical, financial, and organizational risks, the new capabilities would initially augment today's forces; if successful, and as evolved from experience, they would then replace today's capabilities with the more advanced ones that technology will make feasible.

3. The Department of the Navy and the naval forces must change the way they think about building and financing the forces. They must think in terms of life-cycle costs; people, platforms, weapons, and mission subsystems designed together as single systems; and investment in total and enduring capabilities, rather than system acquisition, support, and manning separately. "Affordability" must be thought about in terms of value received for money that is spent within allocated budgets to achieve a desired or necessary capability, rather than as simply spending the least amount of money in any area, as the term has often come to be used.

4. Even this gradual approach will mean a commitment to shifting resources from ongoing programs and operations to new and challenging concepts, and accepting the risk that there will be failures in some cases. There must obviously be a broad base of support for such actions within the Department of the Navy and throughout the Defense Department, the Executive Branch, and the Congress. Without it, the naval forces could not be confident that resources made available by enhancing efficiency or reducing some current capabilities of lesser priority could be retained for application to the desired new capabilities. Building the base of support will be part of the restructuring task.

Preliminary steps toward restructuring the naval forces have already begun, in approaches to using information in warfare, in the emerging Operational Maneuver From the Sea doctrine and concepts of operation, in personnel management, in new and proposed ships, aircraft, submarines, weapons, and their employment and logistic support, and in joint operations and usage. Review of an illustrative example (in Chapter 8) shows that a feasible evolutionary path, accounting for past and current investments in durable systems over their useful service lives, can be followed that will lead to the revolutionary new naval force capabilities that the force restructuring will bring into being. The resulting forces will be more capable and more adaptable to the unexpected challenges of an uncertain future than are today's forces, thus warranting the risks entailed.

The desired future capabilities identified in this study are in the areas of information, people, fleet combat systems, undersea warfare, Marines' combat capability, logistics, and modeling and simulation, with an essential, focused, steadily supported R&D program underlying all of them.

Priorities in creating these capabilities cannot follow hard and fast rules, but rather must reflect a flexible rationale based on progress in crafting the new forces. Priorities may change as programs go through various stages of planning, acquisition, and deployment. In addition, some investments will merit attention simply because technology advances will offer important opportunities for improved effectiveness at modest cost and risk.

The following approach for assigning priorities is suggested:

• First are the technologies that lead to information superiority and more effective use of people. Without the information advantage, the forces will not know precisely where to go, what targets to engage, and how to fight. Technologies for effective use of people must be given priority because it is people, operating increasingly complex and automated weapon and support systems, who fight and win wars, or ensure that wars are deterred, and because the naval forces are especially seeking to make more effective use of people in their resource-constrained environment.

• Next are the weapon systems that constitute the strength of the fighting forces: surface and air systems, undersea systems, and land-combat systems.

• Once the capabilities and needs generated in the previous two areas are

known, related essential logistical support must be provided, since the forces will not be able to operate as visualized if the logistic system is not reengineered to support the new capabilities and modes of operations.
• At a similar level of importance, attention must be paid to modeling and simulation (M&S), which is becoming fundamental to virtually all aspects of major modern enterprises.
• Finally, focused, sustained research and development, similarly prioritized, to support all the above areas is essential—without it, progress in the other areas will be haphazard and difficult to sustain.

FUTURE NAVAL FORCE CAPABILITIES

The entering wedges of naval force capability identified in this study that can lead to the restructured naval forces are described below in the order of the priorities just suggested.

Information in Warfare

The display screen used by the commander, from the CINC to the unit commander, with the information on it, the links to sources of information, the sensors and processing nodes that acquire and develop the information, and the links to weapons and their guidance to targets constitute essential parts of a warfighting system just as much as the ships, aircraft, and combat battalions of the Navy and Marine Corps. Although the quest for information advantage is a factor in all engagements at all force levels, "information superiority" overall must be considered a warfare area analogous to antisubmarine warfare (ASW), antisurface warfare (ASUW), antiair warfare (AAW), strike, and others. The entire area must be treated in an integrated fashion. The ultimate description of system characteristics and the impetus to acquire and modernize the system represented by the commander's display screen must come from the operational forces, as do the requirements for the other warfighting systems.

It is essential to ensure compatibility and interoperability of the naval force systems and other Services', agencies', and countries' systems in the joint and combined information "system of systems" and their essential support for the naval forces, whether the other systems are in space, in the atmosphere, or on land. Therefore, the Navy and Marine Corps must ensure that they are represented in joint forums with the other Services and agencies, and, when relevant, other countries' agencies, at levels that will ensure attention to each Service's information needs. This representation is especially important in the space arena, where the naval forces field few systems but are users of many. There will have to be a cadre of people who remain knowledgeable about space systems to continue the effective liaison that has served the naval forces so well in the past.

In addition, the multiplicity of systems for providing information in warfare has reached a level of importance that requires Navy and Marine Corps personnel with dedicated specialties in information systems and information warfare.[1] Appropriate incentives are needed to ensure that the naval forces find and retain personnel with the high level of ability that the area demands.

Resource constraints and technological opportunity will require adoption of commercially furnished systems for much of the communication and other technology associated with information in warfare. The naval forces, jointly with other forces, must take steps (detailed in the main body of this report and in *Volume 3: Information in Warfare* of this study series) to adapt to using the commercial technology and systems for military purposes, and they must provide the additional protections needed to guarantee the freedom from interference and exploitation that military applications may require. Communications will be critical links in the information-in-warfare system, and no means should be spared to ensure that they cannot be disrupted.

The information-in-warfare system may well become so complex that there will be a serious risk of self-jamming and confusion that could flood users with unneeded information or render necessary information inaccessible when it is needed, or cause dynamic command-and-control instabilities in the system. Means for timely information recovery and information understanding by those who need it will be an essential part of the information-in-warfare system, requiring continuous attention as the system grows. System instabilities will have to be guarded against. Also, doctrine and procedures must be developed for sharing the wealth of information with coalition partners in critical situations where their performance can affect operational success.

People in the Naval Forces

All major naval force systems are being designed to operate with fewer people who have more technical capability at their disposal and more responsibility in using it. This system design trend will require a complete revamping of the naval force personnel system in the years ahead, to improve education and training, to enhance job productivity, to improve the health and medical care of

---

[1] "Information warfare" is a term that has increased in prominence in recent years as the information basis of our society has increased. It includes learning all that can be learned about our opponents, their dispositions, actions, and intentions, in as near to real time as possible; maintaining similar knowledge about our own forces and those of our allies, coalition partners, and neutrals who may affect our operations; and taking any steps necessary to deny such information to our opponents, to confuse them about friendly activities and intentions, and to keep them from exploiting our information activities for their purposes. Information warfare in this sense has always been a feature of warfare. In modern times, however, technology has changed the nature of information warfare significantly, and it continues to do so. Information includes classical intelligence, and information warfare includes classical electronic warfare with electronic countermeasures and counter-countermeasures; their inclusion in the larger aggregation does not imply a diminution of their importance.

naval force personnel (including care of combat casualties), and to retain them in the force longer.

Highly qualified, better educated people will be needed to meet the more demanding technical and operational conditions that future naval force systems and operations will impose. Some of them may be made available by lateral entry of personnel from populations not now in the recruiting pool. Known technology can be applied to speed training and improve job performance. Modern medical technology will make available advanced, technically aided support systems for enhanced health care, casualty treatment, and survival. These technologies are advancing rapidly in the civilian sector, and they are receiving attention at the management and research levels in the naval forces, but they are slow to reach the field. Vigorous and successful investment in these capabilities would lead to a "virtual increase" of considerable magnitude in naval force personnel.

Naval force program and personnel managers are aware that investment in an improved quality of life for Navy and Marine Corps personnel and their families is essential for retention and readiness. Current efforts by all the Services to establish models and quantifiable measures of quality of life will provide a basis for calculating the return on such investments (some examples of such measures are given in Chapter 7). Research and analysis are still needed to extend measures of quality of life per se to valid and useful measures of unit and force performance and cost, to inform investment decisions. Ongoing data collection mechanisms and data analysis capabilities must be embedded in organizations with the responsibility and capability to provide timely decision-making information across the spectrum of Navy and Marine Corps leadership.

Minimizing crew size through effective use of technology is a crucial goal for the future Navy. Success in accomplishing the necessary changes will require exceptionally effective training and exceptional reliability and survivability of the technical systems.

Fleet Combat Systems

Family of Land-attack Missiles

Based on the high responsiveness, rate of fire, and precision of rocket-propelled guided missiles, it is projected that achievable future advances in the missile technology and reduction of their costs will make it possible to greatly enhance the suitability and utility of such missiles for ship- and submarine-launched attack systems. A family of such missiles of different sizes (5-in., 10-in., and 21-in. diameters) for strike, interdiction, and fire support will give the fleet greatly enhanced firepower and surge capability, allowing effective engagement of large numbers of targets of many kinds at various ranges in very short times. With appropriate guidance the missiles could also be used against seaborne targets, and the smaller missiles in the family could be adapted to air launch.[2]

The proliferation of such attack missiles will affect the design of surface ships and submarines, and it will influence how combat aviation is used by the fleet. Because it can have such far-reaching effects, phased introduction of this capability is visualized. The missiles would be developed and used from available and currently planned launch tubes in the early phase. Commitment to major system, doctrine, and force structure changes would follow as the technology (including the anticipated cost reduction) proves itself and the forces gain confidence that the anticipated benefits will be realized.

The Navy's "arsenal ship" initiates and exemplifies the concept of a ship powerfully armed with missiles of the kind described, and others, to be available for the fleet to engage opposing forces pinpointed by the naval forces' joint targeting system. Studies of the tradeoffs between efficiency and effectiveness, on the one hand, and the vulnerability of a large increment of military power embodied in one or a few ships, on the other, are needed to guide decisions about optimal numbers of such ships and of missile launch tubes on each such ship. After experience is gained with such ships, detailed studies of the comparative economics and effectiveness of aircraft- and gun-based systems and the missile-based systems, including consideration of all platforms and weapons in realistic scenarios involving the land, sea, and air forces, will be needed to design the mix of such systems in the overall forces.

Surface Ship and Submarine Design

All future ship and submarine designs will be able to take advantage of fully integrated, distributed sensors, actuators, and automation to minimize crew size and maximize system performance with the smaller crews. It will be possible to retrofit existing ships and submarines with these capabilities as well. A significant start has been made in this direction by the Navy's "smart ship" demonstration. In future ship and submarine designs, and in planning retrofits to the extent feasible, the crew, the logistic support, and integrated damage control will all have to be considered parts of the system from the start, and the entire system designed as a whole.

Additional design features made possible by advancing technology will include:

• Passive signature reduction and capability for signature management in all regimes, for enhanced stealth and survivability;

• Integrated electric power systems and advanced electric drive for more efficient and effective arrangement and use of ship volume;

---

[2] The Panel on Weapons of this study concluded that this family of missiles would be the preferred option, over many other missile, gun, and electromagnetic launcher possibilities, for surface-to-surface fire over the long term.

- Surface ship structures made of composite materials, for reduced signature and maintenance;
- Advanced hull forms to enhance speed, seakeeping, and stealth for surface combatants; and
- Open architectures with modular design to enable more rapid and less expensive maintenance and upgrading of weapon and other ship and submarine systems.

Future tactical submarines will embody much advanced technology, especially in sensors, stealth, power density, and efficiencies attending the development of electric drive and continuing research in nuclear plant design. They will have multimission capability oriented toward support of expeditionary naval force operations. This will include the ability to launch and recover auxiliary vehicles. The submarines will be able to fire large numbers of land-attack missiles from appropriately designed vertical launch systems, and they will need the ability to communicate with the combat information system to enable them to carry out sustained attack missions against targets on land when hostile detection and land-based defenses pose unacceptable risks to the surface fleet or its missions.

Fleet Aviation

Piloted aircraft for attack will continue to be needed in situations requiring the pilot's adaptiveness on the spot, visual target identification, delivery of larger warheads than the land-attack missiles will be able to carry, and sustained campaigns where the prospect of aircraft losses remains low. Defensive counter-air will be able to take advantage of networked, multistatic targeting techniques, enabling longer-range engagements with air-to-air missiles and surface-to-air missiles in the "forward pass" mode and alleviating the predicament, which is expected to persist, that foreign short-range air-to-air missiles will closely match those of the United States in performance. Aircraft providing close air support will add locally to the high volume of surface-launched fire support to help sustain the rapid pace of future ground operations.

New aircraft engine, structures, and flight-control technologies are expected to reduce the weight penalty for the short or vertical takeoff and vertical landing capability of fixed-wing aircraft. Thus, special emphasis on short takeoff and landing (STOL) or short takeoff and vertical landing (STOVL) aircraft capable of flexible operation from a variety of ships and land bases is warranted for the next generation of fixed-wing naval force combat aircraft.

Preservation and enhancement of stealth in aircraft design will continue to be essential. Greater attention will be needed to reducing infrared signatures of aircraft to mitigate the threat of shoulder-fired, infrared (IR)-guided surface-to-air missiles (SAMs) at low altitude and of IR-guided air-to-air missiles in air combat, and there will be technologies to help in this area; the problem will intensify as staring IR arrays are introduced into the weapons. Advanced aerodynamics, microsensor activated controls, and materials permitting higher aircraft engine operating temperatures will offer the opportunity to expand the aircraft flight envelope, while new design and manufacturing technologies are expected to reduce production costs significantly.

There will also be a mix of unmanned aerial vehicles (UAVs) in fleet aviation. At one end of the mix will be high-altitude, long-endurance craft that may operate from carriers or be refueled from them in the air to provide the equivalent of a surveillance satellite in stationary orbit over naval forces at sea. At the other end of the mix, UAVs flown and recovered from carrier decks will be used for targeting opposing ground force elements and for other combat-related applications.

Aerial elements of amphibious operations, including attack helicopters, may be launched from large-deck carriers as well as from amphibious ships. Finally, the carriers will continue to operate ASW airplanes and helicopters, and other aircraft involved in surveillance and logistic support.

Carriers will thus become increasingly versatile as multipurpose air bases at sea. Carrier design can be expected to evolve in diverse ways with the need to operate all the existing and new kinds of naval force aircraft. All of the technology advances in crew reduction, signature management, and lightweight superstructures that will shape the next generation of surface combatants will be applicable to and beneficial for carriers.

Undersea Warfare

Antisubmarine warfare (ASW) research and development (R&D) budgets have been allowed to decline markedly in the post-Cold War years. However, capable and quiet nuclear and nonnuclear submarines are proliferating worldwide, many to nations that may become antagonists. U.S. naval forces will be operating in waters along the littoral in modes that favor the submarine, where detection is difficult and with increased dependence on timely logistic support concentrated at sea. At some point, in less time than it will take the United States to catch up again, hostile submarines in this environment could be in a position to seriously inhibit operational maneuvers from the sea. Attention and funding to a level sufficient for the following tasks will have the greatest payoff for ASW:

- Extending the opportunities for passive detection, by taking advantage of advances in microsensors and fiber optics for very large sensor arrays and advanced computing to perform coherent signal processing;
- Applying the array signal processing mathematics and computing developed thereby to multistatic, active detection and tracking;
- Pursuing multispectrum active and passive nonacoustic sensors in parallel with acoustic sensor development;

- Netting all the fixed, surface, air, and submarine ASW assets in a cooperative engagement mode, and providing the essential tactical communications with submarines, both underwater and on the surface; and
- Improving antisubmarine weapons and counterweapons, with special attention to advanced warheads and performance in adverse littoral environments against sophisticated countermeasures and tactics.

Even with the increasing attention being given to countermine warfare by the naval forces, rapid minefield clearance to protect shipping areas and to facilitate over-the-shore naval force operations remains a difficult problem. Still needed are better means to rapidly focus countermine operations, and means for rapid minefield clearance, especially in the surf and craft landing zones. The former can best be accomplished by attention to intelligence, surveillance, and reconnaissance that will allow mine interdiction, minefield avoidance, and concentration of mine countermeasures (MCM) assets only where mines exist. The Global Positioning System (GPS) aboard all MCM and transiting ships and craft will permit significantly narrower cleared channels. Many small (e.g., 30 tons or less) sea and air MCM platforms supported by a suitable amphibious-type "mother ship," and use of some expendable mine clearance vehicles, can provide rapid mine neutralization and clearance capability organic to amphibious forces, equivalent to as many as 20 MHC-51[3] mine-hunting vessels. Rapid clearance in the surf and craft landing zones prior to a landing can be done by "brute force" methods. The most promising of those is the air delivery of large, precision-emplaced and -detonated explosive charges in an analog to a line charge that creates a channel through surf zone and beach defenses by simply throwing them out of the way if they are not destroyed. Finally, today's mine clearance systems must stand down during night hours. If they were equipped to operate effectively at night, that would in essence double the available MCM capability of expeditionary forces.

These enhancements to the ongoing mine warfare programs can, by the middle of the next decade, bring the naval forces much closer to the much-sought capability for clearing mines rapidly in preparation for an amphibious landing, and for keeping strategic waters mine-free. Their value in the forces would persist for decades longer.

Ground Force Operations in Populated Areas

The recent Naval Studies Board report on the Navy and Marine Corps in regional conflict[4] described many steps that the naval forces should take to be better able to operate in populated areas against various kinds of opposition. This report elaborates on some of them, emphasizing means for conquering pop-

---

[3] Mine hunter, coastal—a class of ship.
[4] Naval Studies Board. 1996. *The Navy and Marine Corps in Regional Conflict in the 21st Century,* National Academy Press, Washington, D.C.

ulated areas without incurring very high friendly and collateral casualties and destruction, and also means for disabling the war-supporting capability of populated areas without occupying them. Much of what is discussed in this report is "in work" in Navy Department and Defense Department programs. Two aspects of such operations especially merit top-level attention:

• Making certain that there is adequate and accurate intelligence preparation to enter unfamiliar foreign areas where the local leaders and tactics could surprise and defeat U.S. forces. This will require some "educated guesses" about where such areas might be, as well as years of advanced preparation of plans and reading-in of potential commanders, along with the willingness to have some of that effort wasted because the need to use it may not arise.
• Extending the techniques and the intelligence preparation to terrorism and other nonconventional means of warfare.

Although these are joint and combined responsibilities that extend beyond the military, they are important for the naval forces because those forces are likely to be first on the spot in many crises.

Logistics

Logistics can be the limiting factor in military force operations at the best of times. The new doctrines and methods for "lean" force operations will increase that risk because they call for reducing dependence on large and usually overstocked forward supply bases in the theater of operations, and increasing reliance on delivery of supplies from their source when needed and as needed. The 1996 regional conflict study referred to above describes in some detail how the logistic system must be reengineered to accomplish this during operational maneuvers from sea to shore and for some period thereafter. The present report extends those observations beyond the immediate area of operations. Key areas for attention and application of modern technology include:

• Providing for distributed, computer-assisted readiness support, moving many support functions from sea to shore in the continental United States (CONUS) or a few forward bases, and taking steps to reduce personnel and use them more efficiently in shore installations and operations, just as is planned for shipboard;
• Ensuring total asset visibility from source to user, to reduce waste through excessive supplies in the system and to speed delivery of supplies;
• Building the system around containerized supply delivery[5] compatible with commercial intermodal transport systems;

---

[5] "Containerized" supply refers to packaging of supplies in containers that are not opened between origin and destination, that are tracked by the use of electronic markers and databases, and that are standardized in size and packing modes. Use of such containers is now standard in commercial intermodal shipping.

• Improving the capability for ship-to-shore transport, especially for movement over the beach, and for "retail delivery" to users beyond the beach; and
• Ensuring compatibility with civilian systems, since they may be called on to help when military capacity runs short.

Munitions constitute a large fraction (on the order of 40 percent) of the wartime logistic load. Shifting much of the strike and fire support from unguided bombs and shells to more frequent use of guided weaponry, and from air-launched to tube-launched weapons, is expected to significantly reduce the time required to defeat large target complexes and is therefore likely to affect ammunition resupply requirements for ships at sea and forces ashore in currently unpredictable ways. An exploration of the potential changes in resupply requirements entailed in the extensive use of precision weapons must be undertaken as part of the planning for the reengineered logistic system.

Logistics and support, in addition to communications, are areas in which commercial services will be used extensively for the foreseeable future.

Modeling and Simulation

Modeling and simulation (M&S) demands attention, support, and participation by the top Department of the Navy command and management levels because it affects every aspect of military force design, equipment, and operation. Although many steps are being taken at lower command levels to manage the growing use of M&S, many critical loose ends remain. The necessary integration of viewpoint and utilization cannot "just happen" without such attention and support. Especially needed is attention to the following:

• Compatibility, consistency, and seamless interfaces between Navy and Marine Corps approaches to using M&S, and inclusion of the implications of the joint environments of expeditionary warfare;
• Coordination of inputs to the Joint Staff Simulation System (JSIMS) and Joint Warfare System (JWARS) simulation programs that will drive much of DOD planning, including that of the naval forces; and
• Ensuring that existing models and simulations are upgraded or, if necessary, replaced (1) to give them a sound theoretical basis in accord with current knowledge and theory describing adaptive behavior of systems and forces in combat and in other aspects of modern warfare; (2) to account for uncertainty in threats and planning; and (3) to incorporate new ways of programming the behavior of networked systems. At present there is little or no empirical support to attest to the credibility of models and simulations used to make major system acquisition and military operational decisions. The M&S community must provide that support as part of an ongoing M&S R&D program by testing their models against real-world events and data wherever feasible. Databases to support such validation efforts must be built.

RESEARCH AND DEVELOPMENT

The section titled "Focused Research and Development" in Chapter 7 of this report, and the eight panel reports that constitute the main output of this study, describe the many areas of research and development needed to create the force capabilities that will shape the naval forces of 2035. Without a strong and sustained R&D program, few of the desired advances will be achieved.

The research and development section in Chapter 7 describes the areas of technology that especially require concentration in Department of the Navy R&D. Within other areas, the preponderance of R&D may be performed by civilian commercial enterprises, so that the Department of the Navy can concentrate its R&D efforts on military applications of a product. However, since civilian enterprises are coming to have a short-term view in today's competitive environment, the Department of the Navy must first ascertain that the civilian world will indeed meet its basic, long-term research needs before giving up such research in any key area.

Aviation has benefited strongly from sustained Navy Department and joint R&D efforts such as the Integrated High Performance Turbine Engine Technology (IHPTET) program. Such programs are especially needed now in surface ship and submarine R&D. These areas are all similar in that they make timely and sometimes rapid progress by steadily building on successive advances in specific technical areas, with periodic application of the advances to a major product development when new levels of capability have been achieved. Other unique areas requiring special Navy efforts in R&D include oceanography and ASW. The needed advances in these and the other areas listed in "Focused Research and Development" in Chapter 7 will be much more difficult to achieve without such a focused, sustained, and coordinated R&D program.

3

Recommendations

Based on the results of this study and on the implications of those results as discussed in Chapter 8 of this report, the following recommendations are conveyed to the Department of the Navy.

Recommendation 1: Plan and implement an aggressive program to create the entering wedges of capability that will position the naval forces to meet the challenges of the 21st century. Key technical capabilities anticipated by this study include:

- Information superiority as an integrated warfare area; capitalizing on and adapting to the vast commercial infrastructure;[1]
- Technological support for highly qualified, better trained, and better educated people, retained in the force longer;
- A family of rocket-propelled, surface- and submarine-launched, land-attack guided missiles (adaptable to air delivery and to other missions);
- In combat aircraft: STOL, STOVL, standoff, and stealth;
- Air-to-air cooperative engagement at long-engagement ranges;
- Stealth and automation in ships, which must be designed as complete systems;
- Unmanned aerial and underwater vehicles providing essential capabilities for combat;
- Greatly expanded capability of submarines to support forces ashore;

---

[1] Of necessity, the information system will include some organic targeting capability as a fallback.

- Advancing ASW through coherent signal processing and cooperative engagement in undersea warfare;
- Becoming able to clear mines rapidly during expeditionary operations;
- Ability of small units to neutralize large, built-up, populated areas with minimal casualties and collateral damage;
- A logistic system based on the use of modern information technology with lift, ships, and processes tailored for supporting forces at sea and ashore from the sea;
- Modeling and simulation applied to acquisition, readiness, deterrence, and warfare: theory and methods to suit the needs of future naval forces for deterrence and warfare.

Recommendation 2: Design, implement, and sustain a vigorous program of naval systems R&D to create the new capabilities.

This program should capitalize as much as possible on commercial technology development, while sustaining Department of the Navy science and technology and advanced development oriented toward specific naval force needs that the commercial world will not meet. The areas to be covered are described in the eight panel reports that present in detail the results of this study.

Part II

Overview Discussion

4

Introduction

The present period appears to the nation to be a relatively quiet time in international affairs. There is no major war under way, U.S. involvement in international quarrels is oriented toward establishing and keeping the peace, and we see no imminent threat to our national well-being and survival. The nation's interest and energies are focused on the development of the civilian society and on the economy and its relationships with economic developments on the international scene.

However, appearances can be deceiving, with ferment below current thresholds of general notice. The past century saw U.S. involvement in two World Wars and the Cold War. Although the size and capability of U.S. and allied forces, and favorable developments in international relations among the major powers, kept the Cold War from erupting into a third World War, it nevertheless included two "hot" wars, in Korea and Vietnam, that were of considerable magnitude if measured by the number of military casualties and civilian deaths, and it was followed immediately by the need to fight another war with large forces, in the Persian Gulf region, when major U.S. interests abroad were threatened. The coming period is fraught with international tensions that carry risks of deterioration in similar directions, if not met resolutely and with appropriate national security forces.

Chapter 5 of this report reviews this ferment and shows that, given the trends in development and diffusion of military technology and capability in the evolving world political scene, this is not a time to be complacent or to let our collective guard down. Rather, it must be considered a time of respite in which we can build to meet challenges that will surely arise as the new century unfolds.

The world of international politics and national security in which the armed forces play a key part often changes much more rapidly than the armed forces can be changed. A great deal of adaptability must therefore be incorporated into the armed forces from the start. Their form, modes of operation, and military capability are driven in large part by the technology available to be incorporated into them as they are built, whether that technology is available from the civilian world or is developed explicitly for the armed forces. Continual review of the nation's strategic situation, the state of the armed forces and the trends in their development, and both their current and projected future suitability to help ensure the nation's security in emerging strategic situations, is thus an essential part of the construction and maintenance of effective armed forces. The study reported on here contributes to such a review for the naval forces—the U.S. Navy and Marine Corps.

At the time of the 1988 Naval Studies Board projections of the naval forces' future,[1] the international political world was already in transition from the world of the Cold War. Nevertheless, it still appeared that the main challenge to the naval forces would come from the Soviet Union, although changing world conditions made it clear that naval force operations in what was then called "the Third World" would become increasingly important.

An update of the 1988 study, published in 1993,[2] recognized that with the shattering of the Soviet Union into constituent states in 1991 and the consequent waning of the military threat that the USSR had posed to the United States and its allies, international political, economic, and military activity in many other quarters of the world would have a growing impact on U.S. national security. It was found that, in general, naval forces' developmental trends were moving in directions appropriate to the changing strategic situation. An important concern expressed at the time was the need for national recognition that the collapse of the Soviet Union had neither eliminated threatening Soviet systems in case of a resurgence of hostility nor obviated other military threats to U.S. national security, and that the strength of the evolving naval forces would have to be sustained over the long term.

The outcome of the Navy-21 update was consistent with the results of the Bottom Up Review (BUR)[3] of all the armed forces' status undertaken at about the same time. After that review, the need to sustain the armed forces' size and readiness to engage in two major regional conflicts nearly simultaneously (based on dangers seen in the Middle East and in Korea) dictated that with defense budgets tightening steadily, force modernization would be slowed significantly.

---

[1] Naval Studies Board. 1988. *Navy-21: Implications of Advancing Technology for Naval Operations in the Twenty-First Century,* National Academy Press, Washington, D.C..

[2] Naval Studies Board. 1993. *Navy-21 Update: Implications of Advancing Technology for Naval Operations in the Twenty-First Century,* National Academy Press, Washington, D.C.

[3] U.S. General Accounting Office. 1995. *Bottom-Up Review: Analysis of Key DOD Assumptions,* NSIAD-95-56, Washington, D.C., January 31.

Now, with a congressionally mandated Quadrennial Defense Review[4] just completed, it appears that the national military strategy might well be adjusted to give higher priority to modernization because, as was recognized from the beginning, needed modernization could not be deferred indefinitely.

Despite the lower priority accorded to modernization after the BUR, the development of the naval forces has not remained static. The Marine Corps has been developing a bold new concept known as Operational Maneuver From the Sea that capitalizes on new capabilities being acquired in aviation, in amphibious ships and landing craft, and in naval fire support.[5] The Navy has developed cooperative engagement capabilities among networked defenses at sea and ashore that greatly strengthen the naval forces' ability to defend both sea and land forces against attack by stealthy aircraft and missiles. The Navy has also originated the concept and begun acquisition of an "arsenal ship" that can be a base for launching missiles against land, sea, and air targets on command from elsewhere in the fleet, in a networked mode similar to that of the cooperative engagement capabilities for defense. Both Services in the naval forces are working on technical and operational measures to reduce the number of personnel in "overhead" activities that support the fighting forces while strengthening the ability of fewer people to provide such support—the Navy in ship and base design; the Marines in logistic support for operational maneuver forces ashore. At the same time, both Services are also increasingly contributing to and becoming embedded in the joint and national intelligence and information networks being built to support expeditionary operations by all the forces in a theater.

Although these developments are impressive, especially in view of the severe financial constraints under which they have been taking place, they have not yet been fully integrated into new concepts of total naval force design, nor are they being supported in a manner or at a level that would enable the rapid evolution of integrated naval forces in keeping with the strategic demands that the future will place on them. Moreover, technology in the civilian world is developing rapidly in many directions, and this technology will affect the evolution of the naval forces. They will have to rely heavily on civilian technology, and will have to devise new ways to meet technological challenges in areas in which we were able to maintain a dominant position during the Cold War era, but in which such a position is no longer assured.

The purpose of the present study is to explore these concerns in depth, to help the naval forces arrive at integrated plans that will best help them meet the nation's potential strategic needs. Much can happen in the 35- to 40-year time period covered by the study, but different aspects of events will occur at differ-

---

[4] Office of the Secretary of Defense. 1997. *Report of the Quadrennial Defense Review*, Washington, D.C., May.

[5] Naval Studies Board. 1996. *The Navy and Marine Corps in Regional Conflict in the 21st Century*, National Academy Press, Washington, D.C.

ent rates, as will the advance of different applications of technology. As we have seen many times during the 20th century, 5 to 10 years can be a long time in the international strategic arena. At the other end of the scale, some of the newer major Navy platforms, such as nuclear-powered aircraft carriers currently under construction, can be expected to have a useful service life extending beyond the 40-year time horizon of the study. Other areas of technical capability will advance at generational turnover rates varying from 1 or 2 years to 1 or 2 decades. *Thus, the study covers perhaps one generational change of major Navy ships; one or two generations of combat and support aircraft and of weapon system technology; several generations of information and computing technology; and several generations of technology and process in training, utilizing, and caring for the personnel of the naval forces.*

This uneven advance of events, forces, and the forces' constituent parts contributes to the difficulty of creating forces for an uncertain future. Conservative planners, unable to foretell the future with any degree of confidence, may be reluctant to give up tried and tested capabilities for new ones with which there has been little or no experience. This has always been true. Nevertheless, advancing technology does lead to distinctly different kinds of naval forces from one generation to the next. In the 40 years from 1865 to 1905 the Navy changed from a force of mainly sailing frigates to one built around dreadnaughts, and from that to one built around carrier aviation in another 40 years. Nuclear submarines and guided missiles for all purposes flourished in the subsequent 4 decades. The ground forces moved from horse cavalry to armor in 40 years, and from armor to heliborne assault in a similar time period.

This study attempts to show what the next generation of integrated naval force capability can become, taking account of the technical and operational risks involved in changing from one kind of system to the next. The study deals with four areas of concern that will be fundamental to Department of the Navy planning for future forces:

- The international security environment over the period 2000 to 2035;
- Technological opportunities during the period 2000 to 2035—technologies that the naval forces can use, consequent changes in the forces' composition, and how newly constituted forces could operate;
- Shaping the naval forces of 2000 to 2035—describing the technology-based capability that must be made available to enable them to meet the challenges of the anticipated environment most effectively, and to hedge against uncertainty; and
- The implications for Department of the Navy force planning and force building.

The remainder of this volume deals in turn with each of these major areas of concern. The reports of the eight separate panels published in an additional eight volumes (see Box P.1 in the Preface) provide the details of the technologies and their applications that are likely to contribute to shaping the naval forces' capabilities over the time period of interest. Included in those reports are detailed recommendations for actions to advance the naval forces' technological and derivative operational capabilities in each of the areas covered. This overview report describes and discusses the overall, integrated approach to force evolution that emerged from the study, and presents the key areas requiring the attention of the Navy Department's top management and commands to ensure effective implementation.

5

The International Security Environment: 2000–2035

INTERNATIONAL SECURITY TRENDS

Many conflicts during the Cold War originated in the Soviet drive to expand communism and Soviet influence throughout the world, and the efforts of the United States and its allies to contain that drive. Complicating factors derived from the struggles attending the disengagement of the European powers from their pre-World War II colonial empires and the playing out of the Chinese communist revolution. These motivations for international conflict and fights within nations have gone, to be replaced by conflicts over resources, ethnicity, and national or regional dominance.

The nature of conflict has also changed. In so-called conventional warfare it has become important to distinguish national governments and leadership—those responsible for initiating wars or crises leading to war—from populations, whom we do not wish to harm physically for humanitarian and political reasons. This leads to a sustained trend toward precision in targeting and weapon delivery, to attack only the war makers and their ability to make war, and to avoid producing casualties and random destruction among local populations. Targets will therefore include not only those that would be on attack lists during any military conflict—critical command, control, communications, and intelligence (C^3I) nodes, transportation hubs, airfields, logistic centers, storage sites for weapons of mass destruction, and fielded forces—but also those that are components of the adversary's civil and governmental infrastructure. The latter include public utilities, telecommunications networks, banking systems, mass media, civil transport, and law enforcement centers—in short, anything that supports the opponent's ability to prosecute modern warfare. To attack the infrastructure targets without indiscriminately harming the civilian population of the area, new kinds of forces, precision weapons, and tailored modes of attack are required.

While warfare between national armed forces or the threat of such warfare continues, the world has also seen the rise of mob violence, guerrilla warfare, and terrorism as a means to disrupt and to destroy established governments. The international drug trade, rife with violence and technical measures and countermeasures, has assumed the dimensions of an international struggle of major proportions. Dependence on computer databases and their intercommunication has opened the possibility of attacks on national and corporate information infrastructures that, if successful, can seriously disorient and weaken the very foundations of modern technological societies.

All of these modes of conflict and threats to the peace cross national boundaries as we have known them, although in many cases national entities aid and abet them. The transnational groups and their mentoring countries also have it in their power to acquire weapons of mass destruction—nuclear weapons, and especially chemical and biological weapons that are difficult to deny to the would-be users and difficult to detect. Small groups and countries can threaten both their neighbors and major nations far from their borders. The developing transnational threat to order and peace in the world is not amenable to solution by traditional diplomatic and military means.

During the Cold War the play of events was subject to the "virtual discipline" of fitting in some way into the major two-sided competition between the free and the communist worlds. As a consequence of the new developments on the international scene, it might be said that we have entered a new, different, and more complex period of cold war characterized by unfocused but incessant world conflict. Many of the key action areas—vis-à-vis terrorism, drugs, and information warfare—are not primary naval force responsibilities, although at various times and places the naval forces must deal with them. They will demand new connections and working relationships among U.S. military, intelligence, and civilian agencies.

The expansion of the transnational threats to our security has received increasing attention in the absence of a major threat of global warfare. We must remain aware, however, that the post-Cold War world is also seeing the gradual emergence of regional national power centers that will be in a position either to threaten or to reinforce regional security and our own interests in various parts of the world.

In the Middle East, beyond Iraq's continuing hostility, the warlike activity of religious fundamentalists in the broad arc from Algeria through Afghanistan, encouraged and in some cases actively aided by Iran, signals a movement in much of the region toward hostility to the United States and its interests that may be hard to counter in the long run. Iran is also building its conventional armed forces and is reaching out for closer relationships with other Muslim countries along Russia's southern boundary from Turkey to Kazakhstan.

Russian economic and political development remains unstable. There could be a resurgence of Russian nationalism and xenophobia, and a consequent resumption of hostile confrontation with the United States and its allies. Or, given enough time and a run of good fortune, Russia could eventually grow into a friendly regional or, again, a world power. Russia's size, its resource base and industrial potential, and its residual stores of nuclear weapons and delivery systems make it a force to be reckoned with, now and in the future.

The Pakistani-Indian conflict over Kashmir could result in another war in the area, with concerns expressed by many in the West that such a war could become nuclear, and therefore damaging to other parts of the world. The tensions generated by that conflict, and by earlier Indian closeness to the Soviet Union, caused "prickly" relationships between the United States and the nations on the South Asian subcontinent. Whether the tensions continue, while India tries to build its industrial and military strength and Pakistan pursues a nuclear deterrent, will depend on many unpredictable events of communication, miscommunication, and perception of real or imagined slights, threats, or assists to regional interests on both sides.

China is growing as an economic and military power, pressing outward and becoming more assertive on the international scene while its government remains intransigently authoritarian. As part of its new assertiveness, it has signaled its interest in exercising sea control to significant distances—up to 2,500 km in some directions—from its coast. And, as we were reminded by a Chinese spokesman during the Taiwan crisis of 1996, China has the ability to launch nuclear-tipped ICBMs against the United States. While China has taken steps to ease tensions along its borders with Russia and India, it has taken a totally independent approach to foreign policy that has been consistently inimical to declared U.S. interests. This has included surface-to-surface missile sales to Pakistan, Iran, and Saudi Arabia, and suspected assistance to Pakistan in the development of nuclear weapons. Clearly, China will be a major force to be dealt with in the coming years and decades.

But other countries on the Pacific Rim will also loom large in future U.S. economic welfare and national security. Korea remains a potential flash point until there is some resolution of the deteriorating position in the North and some move toward peaceful reunification. Japan occupies a position in the Western Pacific not unlike that of Germany in Europe—an economic powerhouse having a xenophobic history that demands continual, friendly engagement on our part, lest some precipitating event cause it to assume an independent course that would surely lead to clashes we would much prefer to avoid. Growing economic power centers in Southeast Asia, such as Indonesia and Thailand, keep a wary eye on both China and Japan. They will be quick to note if the United States weakens its security commitment to the area, and their history suggests that they would then shift orientation accordingly.

The prospective international situation is summarized from the naval forces' point of view in Table 5.1. The table includes specific projections of potential

THE INTERNATIONAL SECURITY ENVIRONMENT: 2000–2035

TABLE 5.1 Future World Scene and Potential Naval Force Operations

Where	Projected Status (aside from unexpected alliances)	Need for U.S. Naval Forces	Likely Kinds of Actions
Northwest and Central Europe; Western Mediterranean north shore	Allied; friendly; economic rivalry	Low	Base area; freedom of sea; combined operations out of region
Russia and environs	Unpredictable—friendly Neutral, prickly ot hostile; economic rivalry later	Low if friendly; medium high otherwise	Any naval force operations possible (except amphibious)
North Africa; Eastern Mediterranean, Persian Gulf, Arabian Sea	Unstable and changeable—some allies, some friendly, some hostile; internal conflicts and transnational terrorism	High	Presence; interposition; resupply; ATBM protection; blockade; could be full range of naval force operations
Indian Ocean	Neutral/friendly;	Low	Presence; freedom of seas
Japan	Friendly/prickly; Economic rivalry	High	Presence; freedom of seas in regional waters
Korea	Allied South, hostile North early; if unified,?	High	Full range of naval force operations
China and Taiwan	Prickly to hostile mainland, friendly island;	High	Presence; full range vs. external threat
Southeast Asia through Australia and New Zealand	Allied through neutral/friendly; economic rivalry; China looms	Medium high	Mainly OOTW; very different in the two areas
Africa south of Sahara; South America except Northern Andes	Neutral/prickly to neutral/friendly	Medium high in Africa; low in South America	Counter-drug and other OOTW; U.S. border security
Northern Andes; Central America, Caribbean	Friendly to neutral/hostile; includes transnational drug cartel	High	

U.S. relationships with nations and situations in specific geographic regions, as they appear in 1997. While the discussion above reviews the likely emergence of major regional powers as seen from current events, Table 5.1 is intended more as a review of the entire world situation that may face the nation and its need for naval forces in the future. Many of the relationships described in the table would appear to face the United States with less serious international security problems than could be entailed in relationships with the major emerging regional power centers, but, as we learned in Korea and Kuwait, less prominent concerns can become major ones very rapidly. The predictions in Table 5.1 may prove to be right or wrong in any particular area of the world. Some of the predictions will come to pass, however, in the sense forecast or in some related way, and other relationships, currently unforeseen, will arise. The key point is that the trends portrayed bespeak an unstable and chaotic international situation, in which some eventualities can be foreseen with clarity, but in which small, unforeseen events can lead to big, unanticipated developments, like alliances between two major adversaries who may have been at odds with each other, or the outbreak of major war, that can affect our security profoundly.

From this review, it appears that in the Mediterranean, the Middle East, and the Far East, the need for a U.S. military presence, and especially a continuing naval force presence, will continue into the indefinite future. The naval force presence needed will be a mixture of friendly engagement, deterrence,[1] and outright military action of many kinds reviewed in due course below. In addition, continuing naval force presence and operations of various kinds can be foreseen in the waters around Europe, Russia, Africa, and Latin America, and possibly in other areas currently unforeseen.

ADDITIONAL FACTORS IN THE ENVIRONMENT

Bases

During the Cold War there was a clear justification for a worldwide U.S. military base structure, and such a structure was built and sustained with the welcome permission of the host nations. Since the end of the Cold War our overseas base posture has shrunk rapidly, out of budgetary and political necessity; the number of overseas installations, many of them aggregated in major base areas, that are occupied and used by the U.S. military has declined nearly 60 percent since 1990, from about 1,700 to 700. The decline has seen the growth of basing constraints that enhance the need for forward forces able to operate independently for significant periods.

Virtually everywhere, U.S. use of bases on foreign soil is now contingent on host government approval of the purposes for which our forces will operate out of the bases. Within that constraint, there is a patchwork of mixed welcomes, in Europe, Japan, and the Arabian Peninsula, that provides but few opportunities for a sustained presence on the ground where we are free to act at will in our own perceived interest.

The naval forces have a strong advantage in this situation because they can sustain operations at large distances from secure bases, maintaining a continuing, visible presence in a coastal zone without intruding on any nation's sovereignty in sensitive situations. Forward movement of naval forces at sea in times of crisis also creates less tension domestically regarding the advisability of U.S.

---

[1] Naval Studies Board. 1997. *Post-Cold War Conflict Deterrence,* National Academy Press, Washington, D.C.

involvement, and the naval forces can quietly leave the scene without creating a political furor. From a forward posture the naval forces can also move rapidly to secure base areas for the other Service forces to move into when needed, against opposition if necessary.

Coalitions

The regional interests of the United States must always involve other nations, on either side of quarrels that may well up. We will almost always have to operate in coalitions, with the consequence that our freedom of action will usually be constrained by competing interests of our coalition partners. Those interests will change, so that the coalitions themselves may change with local situations at any time. Submitting to coalition constraints may not sit well in the U.S. domestic political situation, but it is likely to be a continuing fact of life for the naval forces. The naval forces are, in fact, well positioned to help with coalition building, because they exercise frequently with other nations' navies, even in circumstances that make exercises involving land-based forces too sensitive to pursue. Such exercises help build and sustain readiness in the coalition context, as well.

Resources

The shrinkage of resources for the armed forces means that the tension between maintaining forces of a size and readiness to respond to crises that may arise quite rapidly somewhere in the world, and keeping the forces modern so that they can match or exceed improvements in foreign military capability, will continue indefinitely. Thus, unless one of the regional challenges described above grows into a major military threat of the kind that faced us in the Cold War, the tightness of the defense budget is unlikely to be relaxed. Incorporation of new and advanced military capabilities in the naval forces will have to be undertaken at the expense of forces in being or some other aspect of force posture, by shifting resources within fixed or shrinking budgets. There will be a premium on increasing the efficiency and effectiveness of forces of any size, to render them able to do more with less.

"Jointness"

A convergence of major financially, technically, and operationally driven trends will require that naval forces must increasingly be created and operate jointly with other Service and National[2] agency forces and resources. Extensive

---

[2] The term "National" refers to those systems, resources, and assets controlled by the U.S. government, but not limited to the Department of Defense (DOD).

equipment and mission sharing are implied, as are needs for multiway compatibility and interoperability, task sharing, and information sharing among all the elements of the joint and combined operating and warfighting system.

Technology

Coming out of World War II and early in the Cold War, the advance of technology throughout the U.S. economy was led by defense technology. With the growth of the world's economies, and with the revolution in solid-state circuitry, the armed forces have now come to represent too small a market to dominate the burgeoning commercial markets in many areas, such as computing, commercial aviation, and communications. Thus, the Defense Department and the military Services now depend for much of their technology base on developments in civilian markets. Exclusive military technology developments remain but are limited to specific areas that commercial goods and capabilities have no need to use.

That the United States would have to maintain military technological superiority over our adversaries was an article of faith during the Cold War. A concomitant of the increasing dominance of civilian technology is the spread of the technological capability to much of the rest of the world. That, and our military's increased dependence on the civilian technological base, will make it increasingly difficult for our naval forces to achieve and sustain technological superiority over potential opposing forces.

Moreover, discussion of "revolutions in military affairs" notwithstanding, it must be observed that technological change usually comes in small steps, even though those steps may come often. Even the computer revolution, based as it was on the development of integrated circuits on chips, has taken about 30 years to reach the stage that today we recognize as "revolutionary." This poses the risk that in a tight budget environment it may be tempting for the U.S. naval forces to forego modest changes that have revolutionary potential, while others adopt them to our ultimate detriment.

MILITARY CAPABILITIES OF POTENTIAL ADVERSARIES

One consequence of the spread of advanced technology is the growth of potentially very capable military opposition to any U.S. military operations.[3] Even in situations of lesser conflict or operations short of war, opponents may

---

[3] Much of the following description of opposing military capabilities draws heavily on the Naval Studies Board's report on regional conflict in the 21st century (Naval Studies Board. 1996. *The Navy and Marine Corps in Regional Conflict in the 21st Century,* National Academy Press, Washington, D.C.). The anticipated capabilities described have changed little in the months since the earlier report was published. The discussion of the significance of the opposing capabilities of the future naval forces as they are being considered here is original in the present context and report.

field and be able to use some formidable military equipment and techniques. Such capabilities will be available to any opponent, however crude or advanced.

There will be ready access to information from space-based observations, which may be obtained by sophisticated adversaries launching their own systems, or for others by purchase from any of the space data systems offered for sale in world markets. Such data will have resolutions as small as 1 meter, which will give the observers the ability to track ships of the fleet in most oceans and littoral zones with update intervals of fractions of a day, and to see major elements of ground forces and locate them with respect to known local ground features in a geodetic grid.

Any regular or irregular force may be adept in the use of concealment, cover, and deception, and many have demonstrated exceptional ability to exploit the international news media for their purposes. All will have available capable low-altitude air defenses. These will include shoulder-fired, infrared-guided SAMs of Stinger or subsequent vintage that are very difficult to countermeasure, and advanced, vehicle-mounted antiaircraft machine guns of large caliber with lead computing sights and associated night-viewing devices. All will also have skill with small arms, explosives, and fusing, and all will be able to use diverse land and sea mines.

Actual or potential opponents are very likely to have the knowledge and the ability to use the global information and communication infrastructure for their own internal purposes and for purposes of disruption and "info-terror" against the United States and nations friendly to us. Transnational terrorist and criminal groups have already displayed proficiency in using computers, the Internet, and modern communications media. Acquisition and enhancement of such proficiency are easy, inexpensive, and available on a worldwide scale.

Many potential adversaries will also have broad arrays of modern weapons and military capabilities that are for sale in world markets today and that are being developed by several nations that have had or that have recently acquired advanced technological capability. These are likely to include:

- Modern tanks, combat aircraft with state-of-the-art air-to-air missiles, and artillery.
- Radar-based air defenses, including short-range systems like the French Crotale, medium-altitude systems like the Russian SA-6 and SA-8, and advanced, long-range, high-altitude systems like the SA-10 and SA-12 that may have some counter-stealth and counter-tactical ballistic missile capability.
- Tactical ballistic missiles with ranges from 200 to about 1,500 miles. Within a few years they can come to have advanced guidance systems achieving an accuracy of 50 meters or less, and maneuvering, radiation-seeking, guided warheads. Recent unclassified reports about growing ballistic missile capability outside Europe and the United States show that North Korea, Iran, Iraq, Syria, Libya, and Pakistan, among the smaller nations, have ballistic missiles of Scud vintage and evolutionary advances from that point. Israel and India are develop-

ing missiles with a range of more than 1,000 miles. China has sold long-range missiles to Saudi Arabia and demonstrated their use in the Taiwan Straits in 1996, and North Korea was dissuaded from firing a long-range test missile into the Sea of Japan in 1996. While some tactical and theater ballistic missile (TBM) technologies belong to friendly nations and may be protected under the Missile Technology Control Regime (MTCR), experience thus far has shown that in today's world of spreading technology, leakage to actually or potentially hostile nations will be very difficult, if not impossible, to control. Especially troublesome is the possibility of disabling attacks against ships of the fleet, naval forces ashore, or friendly populations and installations along the littoral by such missiles carrying chemical or biological, if not nuclear, warheads, or even severely damaging conventional submunition warheads. Ballistic missile attacks can be made rapidly and with surprise because the weapon does not require much visible advanced preparation for launch. Under current arms control treaty constraints (discussed in Chapter 7), the United States would not be able to return a strike rapidly at the source of such missiles with a weapon in kind, making a fleet ATBM capability critically important.

- Antiship cruise missiles that either fly at subsonic speed but have stealth characteristics that significantly reduce engagement time, or are supersonic sea-skimmers that present similar difficulties. France and Norway have advertised such missiles for sale, emphasizing their stealth capability; China has transferred such missiles to Iran; Russia has well-developed capability and operational missiles of this kind, which could at some future time be sold to countries that might become hostile to the United States.

- Many means of surveillance and targeting, including space systems, aircraft, and UAVs that may provide some information-gathering capability even in the face of U.S. and allied air superiority.

- An array of sea surface combat capabilities, including surface combatants up to destroyer, cruiser, or even, in the future, aircraft carrier level; and small, fast speedboats that are difficult to sink and that can damage our own surface combatants with missile launches or suicide missions.

- Of special concern to the naval forces are the advanced quiet submarines, some nuclear-powered and some of advanced diesel or air-independent-propulsion design. Unclassified estimates show that among nations that are not currently U.S. allies, Russia has 120 submarines, 77 of them nuclear, and is still building vigorously; China has 70 submarines, 6 of them nuclear, with 2 under construction; North Korea has 40 submarines with possibly 6 more under construction; and 20 other nations have, among them, 74 submarines. Within the last group, there are 30 submarines owned by countries having Indian Ocean coasts—18 Indian, 6 Pakistani, 3 Iranian, and 3 South African. As the recent Iranian acquisition of Russian Kilo-class submarines showed, submarine sales are not always heralded in advance, so that we cannot know how these numbers will change for the smaller, currently hostile countries. Although some of those countries may currently have only rudimentary proficiency in the operation of their submarines, it must be assumed that over the 35- to 40-year time horizon of this study any countries that want to use such capability against U.S. or allied naval forces can learn to do so effectively. This would present a major threat to our ability to initiate and sustain expeditionary military operations along the littoral, especially in view of evolving concepts for such warfare that call for extensive fire support and logistic support from the sea. The implication, also, is that past work on torpedo defense must continue, because (as is indicated in Chapter 7) antisubmarine warfare in the littoral environment will continue to be a difficult process with an uncertain outcome.

• Finally, over the period being considered by the study we may expect and plan for the eventuality that despite the constraints of arms control treaties, there will be a gradual proliferation of nuclear weapons in small numbers, and a more rapid proliferation of chemical and biological weapons, to hostile states, all of which might be associated with some of the delivery systems listed above—especially the tactical ballistic missiles—or with other, covert delivery systems that would be hard to detect in advance.

This listing of military capabilities that the Navy and Marine Corps may meet in future military operations makes it clear that the Services cannot rest complacent. While the above capabilities in many quarters may not appear highly threatening today—either because the likelihood of hostilities appears low or because we believe that the countries that have or are acquiring the capability are not hostile—such conditions could change more rapidly than we could build the capability to counter them. Given the time it takes to field new military systems and to develop new tactics and operational techniques using them, especially in the expected tight budget environment, continuing effort will be necessary to meet the potentially demanding opposition that we can see being fielded today.

STRATEGIC SIGNIFICANCE OF THE FUTURE NAVAL FORCE ENVIRONMENT

The United States in today's world does not perceive an immediate threat to its survival and that of its close allies, such as existed during the Cold War. Rather, we see threats of varying degrees of seriousness to diverse interests, distributed around the world. In our open society, the meaning of events and the appropriate response to them are subject to extensive public argument and, often, delayed response and foreign misinterpretation of both the delay and the response.

The current perception, both at home and abroad, is that when we finally commit military forces in an international crisis our objectives are to minimize casualties, minimize costs, succeed rapidly according to time scales that may be prompted by domestic and political considerations rather than by the needs of the situation, and bring our forces home as soon as possible. While these perceptions may not always be accurate, the perception leads to a pattern of response, as illustrated by the behavior of the Bosnian Serbs prior to the U.S.-led NATO intervention of 1995, that is inimical to our interests and our long-term security.

Our opponents in international activities are concerned with matters of survival or dominance in their local areas, rather than simply with protecting "interests." Their response to possible U.S. use of military force plays on the perceived limitations of our commitment, leading them to plan on longer staying power, and on being able to exact more casualties than they believe we will be willing to tolerate. They use elusiveness, surprise, and deception, and face us with the new kinds of warfare—irregular warfare, terrorism, drugs, and economic and social disruption—that they believe we are not well equipped to handle. They use their pockets of sophisticated military capability, such as mines, antiaircraft weapons, or antiship missiles, to subject us to the "tyranny of the single hit" on major platforms, downing an aircraft or seriously damaging a major combat ship to discourage further presence in an area. They know how to exploit our media by continually posing for the American public the question of whether the price in casualties, dollar costs, and damage, even civilian damage to our opponents, is worth paying for the situation and the gains at hand.

Over the long term, the emerging strategy of major regional challengers must be to enforce their own regional dominance while holding U.S. power at bay outside the domains they wish to dominate. They would intimidate their neighbors while building or acquiring what must now be viewed as today's decisive, strategic weapons: economic power; long-range ballistic missiles with nuclear warheads; modern, quiet submarines; aviation armed with antiship missiles to threaten our naval forces; other weapons of mass destruction; and the capacity for massive but covert disruption of the information systems on which both our civilian economy and military forces depend. These weapons, and the new kinds of warfare listed previously, will ultimately threaten the United States at home—not by invasion, but by disruption and destruction.

Thus, at some point what were threats to U.S. "interests" abroad can turn into threats to the survival of U.S. global power. These challenges will arise differently in different parts of the globe, at different times according to different plays of events. All of the rising regional powers and the hostile transnational organizations will have greatly differing political and military styles rooted in their histories and current circumstances, leading them to emphasize and to use various elements of the "decisive" weapons in different ways. We can expect any or all of these challenges to arise over the next 40 years—nearly half a century.

None of this need imply that our relationships with the emerging power centers will *necessarily* be hostile, although the potential for hostility certainly does now and will in the future exist in varying degrees. But we must heed the lessons of the 20th century, which tell us that we must be one of the "big guys on the block," to paraphrase former Secretary of Defense William Perry, and preferably the biggest, to hold our place at the table, to deter threats and conflicts, and to achieve our strategic objectives, regardless of whether the interaction with the others is friendly or hostile.

The naval forces will be expected to meet—successfully—any of the circumstances and weapons, conventional and unconventional, that international developments will impose. The forces must be adaptable, because the exact nature of any threats they will have to meet or missions they will have to carry out cannot be known a priori. And they must be readily expandable if need be, in a form that will meet major challenges as they develop.

6

Anticipated U.S. Naval Force Capabilities: 2000–2035

The national security strategy[1] of the United States defines the nation's broad national security objectives: to protect the nation against threats to our national security; to promote prosperity at home, in part by enlarging our overseas economic engagement and other friendly interactions; and to encourage the spread of democracy as a means of enhancing the security of the international environment for the United States and our allies. These objectives are to be achieved by several approaches simultaneously in both the civilian and military spheres. The armed forces, including the naval forces, are among the means to be employed. The naval forces themselves will need a clear view of the capabilities they will have available, what the forces will be required to do, and how they will perform those tasks.

THE EMERGING SHAPE OF THE FUTURE NAVAL FORCES

Technologies Available

Naval forces represent a combination of people and machines—from ships and aircraft to microprocessors—that allow and help people to do things that would be impossible without the leverage the machines provide. The nature and capability of machines change with advancing technology, enabling people to

---

[1] The White House. 1994. *National Security Strategy of Engagement and Enlargement,* U.S. Government Printing Office, Washington, D.C., July.

accomplish more, with greater knowledge, precision, and control, as the technology advances. Technology is fundamental to naval force capability.

The technology on which the future naval forces will be based is changing rapidly and expanding explosively in many directions. This study's Panel on Technology identified nine major technology clusters that are transforming all of modern economic and social life, and that will affect naval force capability profoundly. These technology areas are listed in Table 6.1, together with examples of the component technologies within each major cluster. The technologies are described and discussed in detail in *Volume 2: Technology* in this study series.

It is extraordinarily difficult to select from this huge array a few key technologies that may drive naval force development. In addition, technologies may emerge in the next 40 years that are not even conceived of today. As is described below, all of the technologies contribute in some way, in different combinations in various cases, to major force capabilities that could not be developed otherwise. Selection of a few technologies that would have a major impact would perhaps include computing, sensing, and materials technologies that contribute to micro- and nano-technology, including microelectromechanical systems (MEMS), and to the enterprise process technologies. Micro- and nano-technology can be used to form "societies of sensors on a chip" that act like "meta-sensors" and actuators. They will come to underlie all sensitive and accurate information-gathering and system controls, with a broad variety of applications ranging from ASW signal reception and processing to "smart" aircraft skins capable of boundary layer control to enhance lift and reduce drag. The enterprise process technologies enable the economical creation and management of large-scale enterprises and the design, assembly, functional integration, and operation of major systems and "systems of systems." But nearly all the other technologies contribute in various ways to what these few enable, and they contribute to other technical advances, none of which in isolation can generate the naval force capabilities that all of them in synergy can make possible.

Capabilities Enabled by the Technologies

The technologies listed in Table 6.1 are useful to or will affect the naval forces only to the extent of the capabilities they make available. Many of the applications, especially if several related ones are taken together, can lead to breakthroughs in naval force capability. A list of such capabilities would include the following, several of which are elaborated with examples in Table 6.2; the table also shows which of the above technologies contribute, in the main, to the capabilities:

- Information-based conduct of warfare and command, control, communications, computing, intelligence, surveillance, and reconnaissance (C^4ISR);
- More efficient and effective use of naval force personnel: fewer people with more and better technical capabilities at their disposal;

TABLE 6.1 Future Technologies That Will Affect the Naval Forces

Technology Cluster	Examples of Component Technologies
1. Computation	High-performance computing; functional, low-cost computing; microelectronics; systems on a chip (micro- and nano-technology); data storage; digital/analog signal processing; aerodynamic modeling; fluid flow modeling
2. Information and communications technology	Networking; distributed collaboration; software engineering; communications; geospatial information processing; information presentation; human-centered systems; intelligent systems; planning and decision aids; defensive and offensive information warfare
3. Sensors	Electromagnetic (radar, optical—including infrared, visible, and ultraviolet); acoustic (sonar, seismic/vibration); inertial; chemical; biological; nuclear; environmental; time
4. Automation	Unmanned underwater vehicles; unmanned aerial vehicles; robots; navigation; guidance; automatic target recognition; ship subsystem automation
5. Human performance technologies	Communications, information processing, health care, biotechnology and genetics, and cognitive processes, as applied to education and training; operational performance of personnel; health and safety; quality of life
6. Materials	Materials synthesized by computational methods; materials with specifically designed mechanical and physical properties; functionally adaptive materials; structural materials; high-temperature engine materials; specialty materials—superconductive, organic coatings, adhesives, energetic materials
7. Power and propulsion technologies	Electric power: engines and motors; high-temperature superconductivity; pulsed and short-duration power (batteries, flywheels, superconducting magnetic energy storage, explosively driven MHD); energy storage and recovery (rechargeable batteries, fuel cells); and microelectronic power controls and power electronic building blocks (PEBBs). Primary propulsion: gun-tube projectile propulsion; rockets; air-breathing missile propulsion; ship, aircraft, and ground vehicle engines
8. Environmental technologies	Weather modeling and prediction—space, atmosphere, ocean; oceanography and oceanographic modeling. Ship environmental pollution control—waste minimization; shipboard waste processing; hazardous materials handling; noise modification
9. Technologies for enterprise processes	Modeling and simulation; simulation-based system design and acquisition; rapid prototyping; agile manufacturing; logistics management; resource planning; dynamic mission planning; simulated theater of war; systems engineering; cognitive process modeling (all contribute major economic benefits)

ANTICIPATED U.S. NAVAL FORCE CAPABILITIES: 2000-2035   45

TABLE 6.2 Capabilities Enabled by Technologies

Operational Capability	Component Capabilities	Contributing Technologies
Information-based conduct of warfare and C^{4}ISR	Large-scale networking Complete situational awareness Resource and mission planning Targeting Information warfare	Information technologies; sensors; computation; automation; environmental measurement; enterprise processes; geospatial
Efficient and effective use of naval force personnel	Advanced health and casualty care BW/CW detection and counters Distributed training and education System design for smaller crews Longer retention of a more professional force	Human performance; computation; sensors; automation; information enterprise processes
"Smart" systems and "systems of systems"	Extensively instrumented and automated platforms, engine controls, automatically controlled machinery—all leading to more efficient use of personnel Instruments associated with personnel equipment and clothing, enabling people to sense and do more	Computation; information automation; sensors; materials; environment; human performance; geospatial
Unmanned systems	Unmanned aerial vehicles Unmanned underwater vehicles Spacecraft Recoverable unmanned weapon delivery platforms	Information; computation; automation; power and propulsion; materials; sensors
Advanced weapon platforms	Ships, aircraft, submarines Missiles, torpedoes	Power and propulsion; materials; computation; sensors; enterprise processes
Advanced weapon systems	"Smart" detection and guidance Automatic target recognition Multistatic missile, mine, and submarine detection Effective attack on target coordinates First-pass target damage assessment	Sensors; computation; information; automation; materials; environment; enterprise processes
Enhanced survivability of major platforms	Low observables and signature management Absorbent materials, shaping, active cancellation Reduced personnel needs	Computation; materials; automation; sensors; information
Cost reduction in acquisition, sustainability, and logistics	Infrastructure operations: simulation-based design; rapid prototyping, agile manufacturing; lower production costs; efficient logistics management Sustainability: in logistic support; in survival technologies and capabilities	Information; computation; automation; human performance; enterprise processes
Environmental sensing and management	Accurate ocean and weather condition forecasting Clean ships and bases	Information; sensors; environment; geospatial
Modeling and simulation	System design System acquisition Operational planning Realistic training and testing	Information; computation; human performance

- "Smart" systems and "systems of systems"—sailors, platforms, controls, detection and guidance, infrastructure . . . (where "smart" means enhanced capability conferred by sensing, information processing, and electronic, electrical, or electromechanical "force multipliers" that enable more work to be done, more efficiently and effectively);
  - Unmanned systems—some with much autonomy;
  - More capable and efficient weapon platforms, with greater survivability;
  - More accurate, effective, weapon systems;
  - Enhanced survivability of major platforms—by passive and active means;
- Sustainability and "focused logistics"—providing the materiel and services to sustain military operations with minimum waste, at lower overall cost, while presenting a minimum presence of support focus in the theater of operations;
  - Environmental sensing and management; and
- Modeling and simulation—applied to system design and engineering, system acquisition, individual training, force training, force design, mission planning, and almost all other military-related activities.

These are the technology-driven capabilities that will shape the naval forces of the future.

Emerging Picture of the 2035 Naval Forces

On the surface, the future naval forces are likely to appear not radically different from today's forces. They will have ships, aircraft, submarines, a variety of weapon systems, and Marines prepared to move from sea to shore and to fight on the ground and in the air. They will be the products of gradual replacement of huge past, ongoing, and committed near-future investments in systems and people that, to all visible indications, remain effective in meeting the nation's defense needs. However, the forces' operating doctrines and methods, their internal arrangements, and the character of their components may be expected to change radically over the coming decades, so that beneath the obvious surface similarities the naval forces in 2035 will work and be constituted differently from today's forces.

The following picture of 2035 naval forces that can be brought into being emerges from a synthesis of the trends in technology and the environment reviewed above.

- While budget pressures may cause the naval forces in 2035 to have fewer people and platforms than today's forces, they will have a longer reach, be capable of a faster response, and have more firepower per unit than today's forces.
- The forces will have fewer, better educated people, more of whom make the Service their profession, with much more "machine power" at their disposal and more responsibility in the use of that power.

- Technological aids for intelligence and all other information gathering, processing, and use will be central to the forces' doctrine, operations, and tactics. There will be joint networks for information acquisition, management, and dissemination, based on sensors in all media from space through undersea. Raw data and processed information will be transmitted via mixed military and commercial global communications networks, and will be sturdy and secure against interruption and exploitation.

- The Marine Corps will operate in dispersed, highly mobile units from farther out at sea to deeper inland over a broader front, in the mode currently evolving into the concept known as Operational Maneuver From the Sea. They will be provided major fire and logistic support at long range from the sea. They will be skilled in military operations in populated areas, including operations in urban terrain and operations other than war, and will have major capabilities in counter-terrorist operations and in information warfare.

- Rocket-propelled missiles with precision-guided warheads, operating within an integrated targeting-through-damage-assessment combat system, will distribute attack firepower widely through the fleet. There will be several sizes of missiles, able to carry out a variety of fire missions from long-range strike to naval surface fire support of troops ashore. Launched from vertical launch system (VLS) tubes on surface ships and submarines, they will have ranges from about 100 km to the maximum range permitted by treaties covering sea-launched missiles.

- Defensive combat operations and systems, from ship self-defense with close-in weapons through ATBM, will always be networked in cooperative engagement modes that extend from the fleet to cover troops and installations ashore. They will be characterized by multistatic sensing, optimal weapon allocation, and remote release of weapons when appropriate. Defensive counter-air, cruise missile defense, and antisubmarine warfare will all be included in these cooperative engagement capabilities.

- Technology and the necessities of force design in an austere fiscal environment are likely to move carriers, versatile as they are today, even further toward becoming multipurpose air bases at sea. They will operate aircraft for air superiority, direct support of troops in combat, antisubmarine warfare, and mine countermeasures; they will launch amphibious operations; and they will operate unmanned aerial vehicles for surveillance and target acquisition. The last will include such activities as launching smaller aerial vehicles to operate with troops ashore, and refueling unmanned high-altitude, long-endurance surveillance craft to extend their stays in naval force airspace indefinitely. Combat aircraft will include types that are capable of STOVL, able to operate from a large variety of flat-deck ships and shore bases without the need for catapults and arresting gear on ships or long runways on land.

- There will be more varied tactical uses of submarines, including land-attack missions using the family of rocket-propelled guided-missiles described above; ASW; offensive and defensive mine warfare; sea control operations;

launching and recovering special operating forces (SOF); and information gathering and information warfare. The submarines will operate singly and in close coordination with other ships and with forces ashore as major capital ships of the fleet. They will be able to launch and recover unmanned underwater vehicles (UUVs) in support of all these missions. Near-shore operations along the littoral will heighten their need for signature reduction and ability to undertake mine countermeasures while also connecting with the other expeditionary warfare forces.

- As suggested in the preceding paragraphs, there will be extensive use of unmanned platforms—spacecraft, UAVs, UUVs, and unmanned ground vehicles (UGVs), many of them operating in autonomous or semi-autonomous modes, depending on where in the loop the human controllers are placed. Such platforms will be used for surveillance and reconnaissance; support of infantry and artillery by scouting, targeting, and elevating communications relays; electronic warfare and electronic support measures (EW/ESM); ASW; and MCM warfare. Quite likely, other uses will become apparent and will be adopted.

- Logistic support, based on commercial practices and founded in a joint logistics force infrastructure, will be streamlined and more efficient. Support will be provided from the sea until, if it is necessary, a base on land is made fully secure; in either case, supplies will be furnished as needed from a forward supply inventory focused and sequenced to meet troop needs, without the need for a huge and inefficient supply dump to draw from. There will be extensive use of commercial firms for maintenance and support services.

- Finally, not least in importance and covering all aspects of naval force operation, task sharing and mission integration in a joint and combined environment will include all the tasks (outlined in the next section) that the naval forces themselves undertake as their contribution to joint and combined operations in expeditionary warfare, as well as the joint and coalition partners' inputs to naval force operations and security. The latter contributions will include intelligence and surveillance data and processed information inputs derived from spacecraft and high-altitude, long-endurance aerial vehicles; many elements of the communications networks the naval forces will use; delivery of weapons with heavy warheads against targets essential to naval force and overall mission success; boost-phase intercept of tactical ballistic missiles that may threaten the naval forces; and many elements of logistics and base support.

WHAT WILL THE NAVAL FORCES BE REQUIRED TO DO?

Formal mission statements for the military Services change according to contemporary needs.[2] To avoid the attending uncertainties for long-term force

---

[2] Compare U.S. Naval Institute, 1986, *The Maritime Strategy,* Annapolis, Maryland; and Office of the Chief of Naval Operations, 1997, "Forward...From the Sea, The Navy Operational Concept," Washington, D.C., March (available online: http://www.chinfo.navy.mil/navpalib/policy/fromsea/ffseanoc.html).

planning, we can examine what naval forces have actually done throughout history,[3] and project such activities into the future. Their activities have been and will be dictated by geostrategic need, while the means by which those requirements for action are met will depend on the capability that the technology available at the time imparts to the forces.

Examination of historical actions and current uses of naval forces and projection of future need for them in meeting the kinds of challenges outlined previously show that they will be required, at various times and places, to undertake all of the following activities, into the indefinite future:

- Sustaining forward presence as instruments of U.S. foreign policy, and using that forward presence for friendly engagement with the governments and armed forces of allied or neutral countries; for operations other than war, such as surveillance of drug-smuggling routes and protection of refugee relief efforts in hostile environments; and to maintain readiness to respond to international crises;
- Participating in information operations in a multitude of ways, from simply gathering strategically and tactically useful information to observing long-range missile tests impacting the sea, or monitoring hostile transmissions and engaging in other aspects of information warfare;
- Establishing and maintaining blockades to prevent supply and support of hostile powers or forces;
- Deterring and defeating attacks on the United States and our allies, deterring attacks on friendly nations, and, in particular, sustaining a sea-based nuclear deterrent force;
- Projecting national military power through modern expeditionary warfare, including:
    — Conducting strategic movement of troops and supplies;
    — Attacking land and sea targets from the sea;
    — Acquiring advanced bases, landing troops ashore, and subsequently supporting the troops with sea-based firepower and with logistic supply;
    — Dominating local seas and littorals to protect the forward operations and their logistic support, by simple presence or successful combat against opposing forces;
    — Sustained combat at sea and on land, when necessary;
- Ensuring global freedom of the seas, airspace, and space; and
- Operating in joint and combined settings in all these missions.

We may safely project that naval force missions, however they come to be

---

[3] See, e.g., Uhlig, Frank, Jr., U.S. Naval War College, "The Constants of Naval Warfare," a paper prepared for the Panel on Logistics of the present study; Mahan, A.T., 1890, *The Influence of Sea Power on History, 1660-1783,* Boston, Little Brown; Harrington, P., 1994, *Plassey, 1757,* London, Reed International Books, Ltd., Osprey Military Campaign Series; Keegan, John, 1988, *The Price of Admiralty: The Evolution of Naval Warfare,* New York, Viking; and Morrison, S.E., 1963, *The Two Ocean War,* Boston, Little Brown & Co.

documented in the future, will encompass the full range of such actions, as suggested in Table 5.1. The naval forces' ability to remain in place for extended periods without necessarily requiring a presence on shore that challenges sovereignty or political sensitivities at home or abroad enables them to carry out many aspects of such missions simultaneously in various parts of the world, subject only to the constraints imposed by force size, resources, and potential vulnerabilities should hostilities erupt without warning.

HOW WILL THE NAVAL FORCES OPERATE?

Naval force operations are expected to be driven by the need to be much more sparing of resources than during the Cold War, while there will be much less certainty about the nature of specific operations or where they will be required. The forces will operate from forward positions, with a few major, secure bases of prepositioned equipment and supplies to support the combat capability of major—brigade-sized—lead elements of Marine expeditionary forces on short notice. Great economy of force will be required, based on early intelligence that will have to be as reliable and complete as the technology and wisdom of the time allow.

Further, there will be heavy reliance on the acquisition, processing, and dissemination of local, conflict- and environment-related information about opposing, friendly, and neutral forces, permitting situational awareness at all command levels that is as complete and accurate as it will be possible to achieve, in times appropriate to the need. It will be necessary to share much of the information with coalition partners, and to ensure communications compatibility so that their operations can mesh smoothly with those of U.S. naval forces. The forces will have to engage in all aspects of information warfare, offensive and defensive, to deny information to opposing forces while acquiring it for our own forces' use. Assimilation and effective and timely use of the wealth of information available, integration of coalition forces into our own information operations, and defense against information warfare attack will constitute the biggest challenge to successful force operation, because without solving these technical and opposition threats to information superiority, the forces will not be able to operate as effectively as they need to operate against potential opposition. Operations, especially information gathering, processing, and dissemination, will be joint, as will many of the systems operated by the naval forces or for their operational benefit.

Forces can expect to be attacked at greater distances from the shore and in any forward enclaves. Therefore, they will be dispersed, and organizations will have to become flatter to shorten command chains and to give local commanders of smaller and more widely separated force units responsibility and authority for local action under overall force command and control. The emphasis will be on combined arms in mutual support. Operations will be characterized by stealth (both in equipment design and in operational modes), surprise, speed, and precision in attacking opponents. Precision will enable massing of firepower and rapid massing of forces from great distances, at decisive locations and times. Much more naval force logistic support will be based at sea than has been the case in the past. Finally, the ground forces will use novel weapons, systems, and techniques that can mitigate the destruction and high friendly and civilian casualties that go with fighting in populated areas. Such techniques will be designed for use against organized military forces and against irregulars and terrorist and criminal groups that may attempt to undermine or capitalize on Marine operations for their own ends.

This, then, is the vision evoked by the revolution in the making. What must be done to implement it, and how will the necessary capability be developed?

7

Entering Wedges of Capability to Shape the Naval Forces of 2000 to 2035

Because the naval forces are built around major platforms that cost billions of dollars to acquire, the current view of the forces' prospects tends to emphasize the constraints, mainly fiscal constraints, that make it difficult to undertake major new directions of force evolution. However, in the spirit of the ongoing restructuring of all industry and government to enhance our competitiveness on the world scene, this can also be viewed as a time of opportunity for renewal and change to better meet the challenges we face. There are some similarities between the current period and the period before World War II, when a lack of resources and a lack of perceived need by the public and government officials kept the armed forces at a very low level.

Even in that constrained environment the "entering wedges" of essential military capability—prototype bombers and fighters, aircraft carriers, amphibious landing craft, radar, nuclear fission—were there to be fully developed and become the decisive systems in winning the war. So, too, in fiscally constrained times such as these, we can prepare the entering wedges of naval force capability to help the forces meet the challenges and hedge against the uncertainties of the future. In the current case, the beginnings of the key capabilities are already with us. The task is to bring these capabilities into a form and a level of operational competence within the naval forces that enables them to be exercised, used in action, proven, and become the basis for military success by forces in being and for force expansion should that become necessary.

In keeping with this philosophical approach, this study has identified the following entering wedges of capability as the most important for future naval force evolution:

1. Making information systems and operations central to all others;
2. Giving individual sailors and Marines more force-multiplying technical capability, more responsibility, and wider influence on the battlefield and in the battle area;
3. Strengthening the combat fleet by:
— Preparing a family of rocket-propelled attack missiles capable of fast response, a high rate of fire, long range, high accuracy, and low cost;
— Changing surface combatant and submarine designs to use such missiles most effectively, and to capitalize on the technological opportunity to increase efficiency and effectiveness;
— And concurrently, preparing new directions for naval aviation;
4. Expanding the techniques of undersea warfare;
5. Preparing new approaches to operations by military forces in populated areas;
6. Reengineering the logistic system for Operational Maneuver From the Sea (OMFTS);
7. Making modeling and simulation integral to all system acquisition, force preparation, and operational decisions; and
8. Ensuring a focused, sustained research and development program to enable and support all of the other entering wedges of capability.

All of these entering wedges of capability are deemed critically important to shaping future naval forces. With one exception (a research and development program), they are listed in rough order of priority that would be accorded for allocation of resources, although preferably some useful level of resources could be applied to each.

The rationale for the priority order is straightforward.

In the first rank are information and people. Information is first, because without it, the forces will not know where to go, whom to engage, and how to fight. People are next, because it is people, with weapon and support systems at their disposal, who fight and win wars, or ensure that wars are deterred. To help ensure effective use of resources in the resource-constrained environment they face, the naval forces are planning for more effective use of people.

Next in order are the weapon systems that constitute the strength of the fighting forces; this capability includes, on roughly the same level of priority, the surface and air systems, the undersea systems, and the most important parts of the land combat systems that will allow implementation of the full force capability described above. Following—but not much lower in importance because strategy, schedules, and success in military operations are often driven by logistics—is the forces' essential support. Also at this level, attention is needed to modeling and simulation, the technology tool that is basic to the successful creation of all major systems and enterprises today.

Although ensuring focused, sustained levels of research and development is listed last, it in fact undergirds the list—without it, the others cannot be accomplished.

INFORMATION SYSTEMS AND OPERATIONS

The Centrality of Information in Warfare

The naval forces' environment includes U.S. military forces and others that may be allied, friendly, neutral, or antagonistic; military facilities with which the naval forces may have to interact in friendly or hostile fashion; surrounding and intermixed civilian activities and facilities; and factors in the physical environment, such as weather and ocean conditions, that can affect force operations. Information about all these elements is derived from thousands of sources—from human and technical intelligence, space-based observations, sensors deployed by the fleet and by troops ashore and other Services and civilian bodies, and stored or newly generated analyses that can give historical perspective and deeper insights than simple observations alone. The precision and timeliness of the information, used for purposes ranging from devising strategies and tactics to controlling operational force movements to precision targeting for weapon systems, are becoming ever more critical in modern crisis resolution, conflict, and other military operations.

Observation and processing capacities and the ability to communicate the results to multiple users are growing explosively with modern sensing, computing, and communications technologies. Today's military forces exist in a mass of information—an "infosphere"—that is essential to their existence and their effective functioning. All naval force elements must be designed to operate in this information environment. Only if they can capitalize on it to create a complete and accurate picture of their current and projected future situations—more complete, accurate, and timely than their opponents can assemble at any time—can our naval forces, limited in size but with worldwide responsibilities, carry out their tasks effectively.

This means that the information-in-warfare system must be considered and treated as one of the major combat systems, just as are the forces' ships, aircraft, or weapon systems. Indeed, in addition to being an important element of all other combat systems, the information-in-warfare system is the fundamental combat system that integrates and propels all the others. The system design must therefore include the doctrine and the organizational capacity to ensure gathering and distribution of information where and when it is needed. Many of the data sources, and large portions of the communications networks, will be operated by others and therefore will not be under direct naval force control. The naval forces will have to work within this joint system, contribute their own system elements for others' use as well as their own, and integrate their own subsystems at purely Navy and Marine command levels to operate with the entire "system of systems."

The Information-in-Warfare System

Information sources include proliferated sensors in all media—ultraviolet, visible, infrared, and acoustic—as well as radar and electronic intelligence (ELINT) receivers. The sensors are deployed in space, at high and low altitudes in the atmosphere on UAVs and manned aircraft, with the forces on the ground and on the sea, and under the sea—in and over friendly, neutral, or enemy territory. They are fielded and operated by many civilian and military agencies, including, among others:

- The Central Intelligence Agency (CIA),
- The National Security Agency (NSA),
- The National Reconnaissance Office (NRO),
- The Defense Intelligence Agency (DIA),
- The Defense Aerial Reconnaissance Office (DARO),
- The National Image and Mapping Agency (NIMA), and
- Various military Service agencies and elements, including but not limited to those of the naval forces, as well as
- Information sources among our coalition partners, whose inputs must be integrated with those of U.S. agencies in reciprocal arrangements.

All have access to diverse resources and mission responsibilities under national, regional CINC, and local force command.

The flow of information to and from all these sensors must be networked so that data derived from the multiple sources can be correlated in time and space. The information the sensors gather must be processed, analyzed, and distributed to various nodes where it can be used directly or in further analysis to serve various users' specific needs. The networks of sensors and processing nodes must permit adaptive tasking, so that data from sensors and analysis of information can be combined effectively for specific purposes in specific areas, while surveillance is maintained in all other areas of interest. Indeed, surveillance, supported by appropriate processing, must be maintained in areas that *might be* of interest, and there must be alerting mechanisms to indicate when those areas merit attention.

The exchange of information among sensors that is entailed in netting them, and transmission of the raw or processed information to users will require sturdy communications networks that have enormous capacity, in both bandwidth and data rate. Although it is difficult to specify the information transmission capacity needed, because requirements are growing exponentially, two facts about the evolution of future communications technology are essential for the military forces, including the naval forces, to comprehend in planning their communications: (1) civilian communications networks are rapidly surpassing military networks in bandwidth, capacity, and rate of growth; and (2) the attending technological developments will outstrip the evolution of military communications technology, except in specialized areas such as resistance to jamming, the need for special security, and hardening against nuclear effects.

Thus civilian communications technologies, including satellite, terrestrial fiber, and wireless communications, will play a dominant role in future military communications. Given its expanding needs, the military simply will not have the resources to establish parallel nodes and networks of the required capacity, but by adopting and adapting civilian communications technology and networks, the military needs for bandwidth and data rate, whatever they may become, will largely be met.

Using civilian communications technology to meet military needs and integrating it with military communications will not be as easy as connecting with the Internet, however. The most expeditious and economical approach will be to acquire commercial off-the-shelf (COTS) equipment and subsystems for all but specialized applications. In adapting the forces and their procedures to use of COTS equipment and subsystems with their inherent characteristics, the military forces will have to create a seamless integration of terrestrial fiber, satellite, and tactical wireless communications composed of diverse commercial and military subsystems. They will have to ensure the availability of surge capacity and priority access when many of the available communication channels may be taken up with ongoing civilian business. They will have to come to terms with regulatory restrictions affecting civilian as well as military communication system users. They will have to ensure that they cannot be denied service by antagonistic or otherwise unaccommodating subnetwork operators (the military will constitute a relatively small subset of users, not commanding extremely high financial clout in commercial markets). They will have to provide for special needs, such as enabling antijam and low-probability-of-intercept (LPI) communications when large segments of their networks are not under their control; survivability and restoration of service in wartime or after natural disasters; and other problems not yet foreseen. Doctrines will have to be devised, often ad hoc, for integrating coalition partners into our own naval force information communications and information networks. Accommodating coalitions may complicate our own forces' operation and increase their vulnerability, but it will also make available combat forces, intelligence and logistic support, and external political support that may be essential to any ongoing operation. Special needs of military operations will require preparation of accurate maps of potential areas of operation, keyed to the WGS-84 common grid that is being developed. These maps and other information in extensive, militarily relevant geographic databases, such as population distribution or trafficability, will have to be prepared so that they can be accessed by military forces through any communication systems, civilian or military, with appropriate security safeguards, on an as-needed and when-needed basis.

The complexity of the information system and the vast amount of ever-changing information in it at any time eventually will make assimilation of information the primary challenge in system use. Concentration on technologies and other means that aid people in selecting and understanding information—recognition theory, digital agents, network appliances, image analysis, spatial decision support, targeted marketing, and others—will be essential. Also required will be decentralized system operation, much like the civilian World Wide Web, in which information is made available as it becomes available and can be acquired by query when needed by a user without placing rigid demands on user hardware and software system design and operation. Applying this concept to the military information system will present special challenges such as ensuring the timeliness of data; understanding time discrepancies among related data elements that can distort overall situational awareness, and therefore can distort mission planning, execution, and outcome; and alerting diverse users to new inputs that are of direct concern to them and require urgent action.

The "system of systems" that can provide such services will surely grow in size as its architecture evolves in the coming decades. In time it will become large and complex beyond easy comprehension by any one individual, group, or agency, leading to the possibility of unanticipated dynamic command-and-control instabilities that will have to be guarded against, thus making information warfare defense even more critical to reliable system operation. Other dangers include the risk of self-jamming or of confusion if conflicting information arrives from different sources thought to be equally reliable. The latter possibility raises the concern that if information from various sources acts as a sort of forcing function for command decisions, and if the timing of arrival of disparate information from diverse sources is in an unfortunate relationship with decision cycle times, then serious command-and-control instabilities could arise in which maneuvering and firing orders lose coherence and become incompatible with situations on the ground. The result could be malpositioning of forces or failure to move them as and where needed, leading to inability to achieve missions, or even to defeat. There are many historical examples of situations where poor information led to military failure. The risks in a plethora of information, poorly integrated, could be serious, especially with "lean" forces that depend on timely and accurate information and domination of enemy response time lines for military success.

The War for Information Advantage

Objects of observation and surveillance will take steps to disguise or mask their locations, installations, and activities. Beyond that, they will try to take advantage of the known characteristics of our sensors and systems to deceive them, to deny our forces information or to lead them astray or cause them to undertake actions that can lead to their defeat. Growing technical capability in sensors and data processing will permit penetration of concealment, cover, and deception, however. Concealment and cover may be penetrated by airborne hyperspectral and foliage-penetrating radar sensors, and deception can be detected through sufficiently sophisticated processing of information from diverse sources with sufficiently powerful computers.

Of even greater concern than the operational problems in mixing and using civilian facilities with the purely military communications, sensors, and computing networks will be defending against deliberate information warfare attack— reading, disrupting, confusing, and denying reliable information and the successful use of information-based systems, or planting false information that remains undetected. Such defense will need many components to create a balanced defense in keeping with the complexity of the information system. A balanced defense will include various steps to deny visibility into military and naval force use of the system; operation with concurrent backup always in place; preparation for degraded operations; and continuous monitoring, auditing, application of protective measures, and active defense against penetration. Perversely, there may be some safety in open use of multiple networks accessible to many users, some of whom will be opponents.

Electronic warfare, including ELINT, jamming of sensors, communications, and navigation systems including GPS, and steps to counter the jammers, as well as possible use of high-powered microwaves to destroy electronic circuits and defense against such weapons, will continue to be part of the war for information advantage. Every combat and support system design will have to account for vulnerability to electronic countermeasures, and will have to provide for counter-countermeasures. Electronic countermeasures will also have to be part of the offensive "kit of tools" that helps weapon systems and forces reach and attack their objectives against effective defenses in order to deny sensor information to the opposition.

Stealth and signature management in ships and aircraft will continue to be essential in denying unit and force movement and targeting information to an opponent. Even where it is difficult to reduce a platform signature to extremely low levels, some signature reduction will help other electronic warfare components to deny information about the platforms to the opposition.

The quest for information superiority is becoming so broad and complex that the aggregation of the various areas involved, such as surveillance, intelligence gathering, defense against information warfare, non-weapon-specific electronic warfare, and others, must be considered a major warfare area in its own right, of status comparable to ASW, ASUW, AAW, and the other recognized warfare areas. The implications of such a change in viewpoint are addressed below.

Instituting the Information-in-Warfare System

It is clear from the above discussion that the commander's display screen—whether the commander is a CINC, a ship or battalion commander, or at some other level of unit command—together with the information on it, the links to sources of information, the sensors and processing nodes that acquire and develop the information, and the links to weapons and their guidance to targets, constitutes a warfighting system just as much as the ships, aircraft, and combat battalions of the Navy and Marine Corps. It is the operative "meta-system" of the *information superiority* warfare area. This system must be acquired and integrated into the forces by the same processes that govern the acquisition and integration of all the other major warfighting systems, with similar, integrated attention at the same command and executive levels in the Services and the Department of the Navy.

In particular, the information-in-warfare system must be managed in an integrated fashion, and the ultimate statement of requirements for the system, the descriptions of its characteristics, and the impetus to acquire and modernize it must come from the operational forces, as do the requirements for and characteristics of the other warfighting systems.

Some of the information sensors and processing nodes, as well as support systems such as GPS, are outside the Navy and Marine Corps, in other Service, Defense Agency, and National systems, including space systems. In these cases, compatibility and interoperability of the naval force systems and other systems must be ensured, and the naval forces must be assured that they will receive the needed utility from the systems. The originators and operators of the other systems, whether the systems are in space, in the atmosphere, or on land, must be kept aware of Navy and the Marine Corps information and information support needs, and the Navy and Marine Corps must be represented in joint forums with the other agencies at levels that would ensure attention to their needs. Department of the Navy senior leadership must be actively involved in this process. Future technology advances and fiscal constraints will heighten the need.

Specific Navy Department attention at high levels is especially needed in the area of space systems, where the Navy and Marine Corps field few systems but rely critically on many. They depend on space systems for environmental (weather and ocean condition) forecasting, navigation, communication, surveillance, reconnaissance, targeting, position fixing, and weapon guidance. In the past, they have been served well by systems that other Services and agencies have fielded with the requirements of the naval forces as well as other requirements in view. The effective liaison between the naval forces and the other Services and agencies that made this possible could come under severe pressure in the future in a fiscally constrained environment unless explicit attention is given to ensuring that the effective liaison continues.

Finally, the information-in-warfare system has reached a level of importance that requires—like the ship, submarine, gunnery, aviation, infantry, and other operational communities before it—information operations and information warfare specialties in the Navy and the Marine Corps. Only with the attending incentives will the naval forces be assured of finding and retaining personnel with the high level of performance and capability the area demands.

ENHANCING THE CAPABILITIES OF INDIVIDUAL SAILORS AND MARINES

Using People Effectively

Because, short of a dire emergency threatening U.S. survival, we will continue to have volunteer armed forces, the naval forces will have to compete with the civilian economy for personnel. Among the many factors in this competition are compensation, the need to provide work and living experiences that will encourage personnel to make Service-oriented career decisions, and—different from most careers in the civilian world—the fact that armed forces' personnel will, at uncertain times, be asked to risk their lives, and consequently the welfare of their families, as part of their jobs. The future personnel pool will include both male and female sailors and Marines, who will come from a rich variety of cultural and educational backgrounds to which the recruiting and training systems will have to be sensitive and adaptive. At the same time, training technology and techniques are changing rapidly, in parallel with the technological evolution of the naval forces' hardware systems. The personnel system of the naval forces thus faces the prospect of complete revamping in the years ahead.

All major naval force systems are being designed to operate with fewer people who have more technical capability at their disposal. Technology, in the form of elaborate, networked instrumentation, automated controls, and integrated information, communication, and transportation systems that can generate fast response by forces far from crisis areas, is being used to streamline and consolidate functions at sea and to move many traditional shipboard maintenance and support functions ashore. The functions affected will vary from shipboard damage control and system maintenance to target acquisition and weapon firing, all of which will be performed with fewer personnel in future naval systems. Manning[1] major combat systems so as to optimize the mix of equipment and personnel thus becomes a parameter to be considered early in the system design process, along with the technical elements of a system. Gone are the days when major platforms such as ships could be built under the assumption that crew would be found to perform whatever functions were needed; personnel

---

[1] The term "manning" is used as a convenient, generic shorthand for assigning personnel, male or female, to organizational and technical tasks within major systems and support bases.

must now be considered to be integral parts of the overall system, from its inception.

As a result, future force design will encourage more naval forces personnel to make the Service their profession.[2] They will require more training and advanced education, and they will carry more responsibility for system operation, both in the narrow sense of making the system work, and in the broader sense of using the system in combat. In terms of both economics and force effectiveness, it will be important to keep the people in the forces longer. There will be advantages in recruiting people who are, on average, better qualified than today's recruits. Recruiting may have to tap people in the personnel pool, such as community college graduates or individuals in mid-career, who are not generally approached in recruiting today. Naval force personnel will thus become more expensive to recruit, train, and retain, with added expense for accommodating their outside responsibilities. These higher unit costs for personnel will offset the savings from technology-based reductions in personnel, putting a premium on achieving maximum productivity from the force.

Aside from using technology to help fewer personnel operate major systems, thus making the assigned personnel inherently more productive, known technology can be applied to speed training and improve job performance, and the naval forces must move ahead rapidly to capitalize on technology for such activities as training in synthetic environments and using simulators to represent parts of systems, thereby shortening the time required for more expensive training with actual systems and forces in their real environment. Computer and communication networks allow distributed training, so that one expert instructor can train people simultaneously at widely dispersed locations, with a consequent reduction in travel costs and time away from assigned stations. Such training can be designed to be adaptive, allowing each trainee to go at his or her own rate, without having to conform to a fixed schedule based on some average training performance; it will also be possible to change the "courseware" easily to fit the different backgrounds of trainees and different circumstances of training and variations in system design. Distributed networks providing access to remotely located experts—e.g., the designers of a system, or the nation's best electronics warfare experts—together with technical aids such as computer-based plug-in diagnostic tools, can help personnel at sea or at far-flung bases to accomplish their tasks expeditiously and effectively without extensive in situ backup.

Quality of Life

Quality of life—the large complex of factors attending job satisfaction and living in the job-associated environment—is a key factor affecting personnel

---

[2] This should not be construed to mean that a military class in our society is being advocated. That is, indeed, an outcome to be guarded against, perhaps by ensuring that the Service professional's family life is rooted in the civilian community.

productivity and retention. Military careers must be competitive with careers in the civilian economy, and expectations for quality of life are now higher than they may have been in the past. The quality of life for naval force personnel depends not only on pay, which is a key factor, but also on the perception that the Services have policies that value and support their personnel, that the Services' leadership takes those policies seriously and implements them effectively, and that the public approves of and supports the missions and values of the people in the Services.

Research confirms what could be understood intuitively, that people feel more satisfied with their lives if they are satisfied with their jobs, and that good matching of people to jobs leads to better job performance and greater job satisfaction. The well-known psychological tests and associated techniques for accomplishing such matching must be considered part of the Service personnel management's kit of tools. Technology can also improve the work environment in many ways, from enhancing creature comfort to providing adequate and suitable tools and machinery to get jobs done.

Deployed sailors and Marines also perform better if they know that during their absences their families are well provided for in terms of housing, schooling, religious and medical care, work opportunities for spouses and older children, and all the other tangible and intangible factors that lead families to feel satisfaction or dissatisfaction with their daily lives. Sailors and Marines also want to keep in touch with their families, a need that is possible to meet with today's worldwide communications networks. But shipboard and remote base policies and routines must provide for it, and must do so without compromising ship or base security. A ship that is operating in emission control (EMCON), for example, cannot allow calls out, nor can it allow tracking of the ship to locate it for incoming cell-phone calls arriving by satellite. Thus, technical means must be devised to support a policy of keeping in touch with families without jeopardizing the force or its operations.

There is also evidence that deployed personnel feel they are being benefited if they can use their spare time to advance their education and technical skills, leading either to more rapid promotion or better job prospects on leaving the Service. Providing such benefits could help make longer stays at sea acceptable to more sailors, thereby reducing the number of personnel ashore who must be retained for rotation. The balance is a delicate one to achieve, since the rotation policy will affect family interests as well. More must be learned about attitudes and interests that affect views of the rotation policy in relation to perceptions of quality of life in the Service.

It is clear that costs will be incurred in ensuring a quality of life that will encourage retention of personnel. The amount of the investments needed to improve living and working conditions must be known and planned for, and ways of measuring their success must be determined. Ongoing research in all the Services is suggesting quantitative measures of quality of life that may be a starting point for obtaining the information needed to assess the effects of such investments.

Many current measures of quality of life (QOL) are based on subjective ratings of such factors as satisfaction with housing, sense of connection to units and communities, and job satisfaction. More objective measures may include retention rates as they vary with compensation and their correlation with subjective measures, use of family services, and other indicators of satisfaction or dissatisfaction with job and family circumstances. Such measures of QOL must be related analytically to broader decision measures that allow reliable estimation of the potential effects of investments in improving QOL on unit and force readiness and performance. Continuing research to achieve this capability will require ongoing data collection and analysis carried out by responsible organizations that will be able to provide timely information to those who will make the investment decisions and oversee their implementation. The research must also give attention to tracking the results of the investments, to enable continual evaluation and refinement of QOL-related actions.

Caring for Naval Force Personnel

Because the sailors and Marines of the naval forces are ultimately there to fight if need be, some of them will become casualties, of combat or of the exotic environments in which they will operate, and they must be appropriately cared for. Modern technology will allow naval force personnel to be embedded in advanced, technically aided support systems for enhanced survivability.

Whether personnel are at home or deployed, in combat or noncombat conditions, more casualties can be expected from sickness and disease than from combat or high-risk operations. Modern trends in medical care emphasize maintaining "wellness" rather than simply treating those who are ill. This approach covers the gamut from preventing disease to encouraging healthy living habits, about which more comes to be known yearly. Maintaining health could in the coming decades involve such advanced techniques as gene testing and tailoring work and living patterns to avoid exposure of individuals susceptible to specific diseases and injuries.

Modern wound treatment techniques stress the importance of reaching the wounded soldier or sailor quickly to diagnose the exact nature and location of a wound or injury and initiate treatment within the first half hour or less; success in this step can increase the survival of battle casualties manyfold. Advancing medical technology can provide for rapid treatment in situ to stanch blood loss, support broken bones, and prevent infection. It can provide "artificial skin" for rapid sealing and treatment of burns. It can provide robotic assistance for rapid retrieval and evacuation of casualties, multiplying the ability of a few corpsmen to treat more people in a shorter time. Growing capabilities in telemedicine—the ability of corpsmen or nonspecialist doctors in field conditions to reach experts anywhere in the world for advice and instruction in treating difficult wounds, injuries, or diseases—will further multiply the capabilities of medical personnel in the field.

Detection of attacks by chemical or biological weapons, and use of vaccines or antidotes against their effects, must be an important part of keeping sailors and Marines healthy and fit, and of treating them if they become casualties. Preparations for such attacks involve provision of protective clothing, which must be much improved over current chemical warfare protective gear that greatly reduces the ability of the wearer to perform useful tasks and that rapidly induces heat prostration; extensive use of sensors to detect attack, even down to the individual suit level; and suitable sealing and flow control of ship and combat vehicle ventilation systems.

In all these ways, advancing medical and related technology can lead to healthier naval force personnel and greater recovery rates among casualties. In the long run, the result is a "virtual increase" in force size, with a greater fraction of the precious personnel resource being on the job and productive rather than off the job due to sickness or injury. The naval forces should waste no time in assessing the tradeoffs and taking advantage of the opportunities that rapidly advancing medical science and technology are offering.

THE COMBAT FLEET

The combat fleet consists of the platforms that convey combat power to the locations where it is needed and the weapons that deliver that combat power against opposition targets. The weapons strongly influence the design of the platforms. This examination of potential technological progress in the combat fleet first considers a potential weapon capability that can contribute strongly to future fleet combat strength. It then examines the design of ships, aircraft, and submarines that will use those and other weapons. In closing this discussion of the combat fleet, some issues in deciding directions of the evolution of the fleet are noted.

A Family of Land-attack Missiles for the Fleet

Today the Navy is beginning work on a new kind of ship, the "arsenal ship," so called because its sole purpose will be to carry and launch on command a variety of missiles against targets located and identified by the off-board combat information system. One of the kinds of missiles the ship will be able to launch will be a "marinized" version of the Army Tactical Missile System (ATACMS, or NTACMS in the naval version), a rocket-propelled guided missile with ranges from 100 to 200 miles, depending on the warhead weight carried. This Navy version of the tactical missile will also be capable of launch by any other surface ship that has a VLS or by submarines similarly equipped. The Navy is also developing an extended-range guided munition (ERGM), a guided, gun-launched shell with a rocket motor, to be fired from shipboard guns to ranges of 60 or 70 miles in support of joint and combined forces in action ashore. The ERGM is essentially a rocket-propelled tactical missile whose first stage is a naval gun barrel.

The value of such missiles for strike warfare, interdiction, and naval surface fire support is their unprecedentedly fast response time to target and the ability to launch many weapons simultaneously or in rapid succession to achieve high volumes and rates of fire—sometimes as much as one or two orders of magnitude greater in these respects than warheads conveyed by aerodynamic vehicles. These qualities would enable them to be highly responsive to the needs of the joint and combined forces that have been landed and are operating ashore in the mode that will emerge from the OMFTS doctrine using techniques described in the regional conflict study.[3] They will be especially useful in providing surge firepower at the opening of a conflict or campaign, and for early support of ground forces from the sea. And, with a fleet stretched thin, a small force of surface combatants and submarines can promise heavy firepower in a short time, for deterrence purposes, to buy time for arrival of reinforcements, to fix opposing forces in place, or to destroy them, as the situation may require. Such missiles are also much more difficult to defend against, increasing the assurance of penetration to substantive targets without the need to undertake costly campaigns for suppression of enemy air defenses.

Based on projected advances in rocket-propelled guided missile capability discussed below, a family of three such missiles, defined by diameters of 5 inches, 10 inches, and 21 inches, can be visualized that will meet mission requirements ranging from naval surface fire support of forces ashore to long-range strike of theater-strategic targets. Approximate characteristics of the missiles are shown in Table 7.1.[4]

Especially noteworthy is the large number of smaller missiles that can be carried in standard VLS missile bays.[5] For comparison with these numbers, a

---

[3] Naval Studies Board. 1996. *The Navy and Marine Corps in Regional Conflict in the 21st Century,* National Academy Press, Washington, D.C., pp. 36-39.

[4] The land-attack capabilities of the missiles are emphasized here in keeping with the current power projection orientation of naval forces. It is apparent that with appropriate guidance system adaptation, the missiles could also be used in antisurface ship warfare. It is also apparent that, although this presentation emphasizes the launch of these missiles from surface ships and submarines, the basic designs can be adapted for air launch as well.

[5] The smaller numbers of 5-in. and 10-in. missiles shown in Table 7.1 assume single-stacking of the missiles, with four 10-in. missiles per cell and sixteen 5-in. missiles per cell. A precedent for multiple missiles per cell exists in current plans to stack four 10-in.-diameter Evolved Sea Sparrow (ESSM) air defense missiles per cell. If the smaller missiles could be double-stacked vertically for

TABLE 7.1 Approximate Characteristics of Family of Land-attack Missiles

Mission	Length	W'hd Weight	Range	Number in 64-Tube VLS Bay
5-in. Naval forces fire support	5 ft to 7 ft	50 lb	100 km	1,024 to 1,536*
10-in. Interdiction	10 ft	100 lb	240 km	256 to 384*
21-in. Strike	21 ft	400 lb	> 600 km†	64

*Depending on how they are stacked; see text note 5.
†Range limited by arms control treaties. The START I treaty limits ballistic missiles launched from surface ships to a 600-km range. The range of submarine-launched ballistic missiles, as these missiles would be defined under the treaty, is not limited, but the number of launchers is. The issue to be resolved in separate understandings that might not be reached until posed by the advent of the systems, is whether submarine launchers for these tactical missiles would fall within the treaty launcher limits. The START I range limits on surface-launched missiles expire in 2006, after which renewal or renegotiation would be required. It could well take up to or beyond 2006 to develop and start to field VLS-launched missiles having the range, with desired payload weight, that would raise the issue.

DDG-51 carries around 500 5-in. shells in its magazine, and a DD-963 and CG-47 each carry around 1,000; numbers of ERGMs would be fewer. The missile launch mechanism on the ship, even for multiple-stacked missiles, would be simpler and less expensive than the combination of gun, recoil, and loading mechanisms for high-rate-of-fire guns (according to a briefing by the VLS Program Office, a 64-cell VLS bay costs $2.5 million, plus installation, while other Navy figures indicate that a 5-in. 54-caliber automatic naval gun with a loading mechanism costs $12 million to $13 million plus installation). Use of VLS missile launchers would thus permit more efficient use of valuable shipboard volume.

At this stage of development, missiles such as those described are insufficiently accurate (e.g., they can achieve a circular error of probability (CEP) of less than 20 meters) to compensate for the lower unitary warhead weight the missiles can deliver at long range relative to attack aircraft. Their accuracy is

---

launching, with one group on top of the other, the lower number shown in Table 7.1 would double. A critical problem, however, would be to vent the exhaust gases of the missiles in the upper stack so that they would not damage those in the lower stack. The problem can be solved by appropriate engineering design. To be conservative, it was assumed in calculating the larger numbers of missiles shown in Table 7.1 that the volume required for venting the exhaust gases would reduce the number of cells in the standard 64-cell bay to 48. The constraints on multiple stacking would not apply to missiles cold-launched from submarines. Cold launch can be applied to surface ships as long as the engineering provision embedded in the guidance system is engineered so as to prevent the missiles from falling back on the ship in case of ignition failure. With cold launch and triple stacking, as many as 3,072 5-in. missiles could be loaded in a 64-tube VLS bay.

sufficient for delivery of distributed antipersonnel, antimateriel, and antiarmor submunitions. With increased accuracies achievable at low cost during the next 30 to 40 years through the use of GPS/inertial guidance packages currently under development, the lighter warheads deliverable at the longer missile ranges, with their explosive energy augmented by the warhead's kinetic energy on impact, will be able to deliver destructive energies on target that are sufficient for a large fraction of target engagements. (For example, to the first order, a 400-lb missile warhead striking a target at Mach 5 will deliver total energy roughly equivalent to that of a 1,000-lb bomb striking at Mach 1, although the destructive effects of the energy may be distributed differently.[6]) In addition, higher energy density explosives that are currently a topic of research will, if sensitivity problems can be solved, make the smaller missile warheads much more powerful—perhaps up to a factor of 2 or 3.

The use of a common GPS grid for target location and weapon guidance will reduce warhead delivery error for fixed or "theater-strategic" targets, and it will allow reduction of the target acquisition "basket" for attack of targets that can move during the missile's flight time (as long as they do not move beyond the missile's kinematic target-tracking capability.)[7] The ability to track the latter targets accurately, in real time, that the evolving information system is expected to achieve will allow continual in-flight update of target location using relatively inexpensive one-way data links to the incoming warheads.[8] Or, simple seekers might be used, with target acquisition aided by accurate placement of the missile within a narrow "basket." Such relatively low-cost techniques will ultimately enable the missiles to achieve CEPs smaller than critical target dimensions in many circumstances, and to attack moving targets as well as stationary ones. The energy requirements for target destruction or total disablement will be reduced correspondingly, especially with "smart" targeting—targeting the critical points that will permanently disable the functioning of larger target complexes. Such accuracies will also greatly reduce incidental or collateral civilian damage in target areas, and will allow friendly forces to bring supporting fire much closer to their positions than purely ballistic air- or gun-delivered weapon trajectories have allowed.

Advances in rocket motor design for the land-attack missiles over the 2000 to 2035 time period can include staging and increasing the specific impulse of the propellant by perhaps 20 to 30 percent. Such improvements will increase their range to values well beyond that achieved with rocketry to date. Other

---

[6] A full analysis of target, missile, and warhead interactions would be necessary to match warheads, target types, and modes of attack for these new systems.

[7] Naval Studies Board. 1996. *The Navy and Marine Corps in Regional Conflict in the 21st Century,* National Academy Press, Washington, D.C., p. 51. The concept of targeting coordinates will also require the support of a robust mapping, charting, and geodesy effort.

[8] Naval Studies Board. 1996. *The Navy and Marine Corps in Regional Conflict in the 21st Century,* National Academy Press, Washington, D.C., pp. 63-64.

improvements could include cold or nearly cold launch, imposing less heat and erosion load on the vertical launch tubes and permitting the use of the same missiles in surface ships and submarines. This step would reduce overall system cost. We can also expect further unit missile cost reduction through assiduous effort over the 35- to 40-year time period, by such means as simplification and standardization of guidance, control, warhead, and rocket motor components (even at the expense of some penalty in performance gains attending subsystem simplification) and large-scale production of the resulting weapons (tens of thousands of missiles of all sizes, with many common components).[9]

A New Generation of Navy Surface Combatants

Advancing technology, and the need to accommodate weapon systems such as the above family of missiles, can be expected to lead to many design advances in the next generation of Navy surface ships and submarines. Surface combatant design is discussed here; submarine design is discussed below.

Advanced surface combatant designs will incorporate and extend many features currently in experimentation on today's ships. Indeed, once these features are developed to the application stage, many of them can be retrofitted to greater or lesser extent in today's ships at major maintenance and overhaul milestones during their service lives to increase survivability, system reliability, and ship service life. These features include changes in how the ship's crew is assigned, how instrumentation is integrated to let fewer people operate the ship, and other system changes on the "smart ship," the cruiser USS *Yorktown*. Foremost among the changes in ship design that advancing technology will permit and encourage will be the following:

- Fully integrated instrumentation and automation in design of ships, using distributed and networked sensors, actuators, and microprocessor controls to minimize crew size and maximize efficiency. This will include damage control, a very sensitive area that is currently a major determinant of crew size. Automation in damage control is also the subject of current Navy research, and will be advanced by fully integrating instrumentation, automation, and revised crew functions into new ship design from the start.

- Passive and active signature reduction and capability for signature management in all aspects—wake reduction, noise reduction, hull and superstructure shaping, and electromagnetic and infrared emission control. Even if surface ship signatures remain relatively high in terms of gross detectability in all but a few specialized cases, for a variety of reasons, significant reduction from current values, which will be made feasible at low cost by design and structural changes,

---

[9] Naval Studies Board. 1996. *The Navy and Marine Corps in Regional Conflict in the 21st Century*, National Academy Press, Washington, D.C., pp. 63-67.

will help acoustic and electronic countermeasures function much more effectively to mask ships under combat conditions or in other circumstances where detection and tracking must be made difficult for opponents.

- Open architectures that will allow modular replacement of diverse subsystems whose technology matures at different rates, and to allow off-board maintenance of complex systems.

- Modular design of weapon systems, including plug-in data buses for actuation and compatible containers for the weapons themselves, allowing field flexibility in reprogramming for various weapon types without extensive ship modifications and crew retraining preceding each choice.

- Integrated electric power systems and electric drive, including the introduction of high-temperature superconductivity when the technology evolves appropriately, will enhance volume flexibility in ship design, will improve overall system efficiency, and will help with active and passive signature reduction (although, as with any change of technology, new signatures may be created). Readily controlled electric drive will be enabled by solid-state electronic controls (i.e., through the use of PEBBs) that are now appearing, which will replace bulky switches, transformers, and banks of condensers, and will allow easy management of voltages to different ship systems and AC/DC conversion.

- Ship structures made of composite materials will enable embedded and conformal sensors, specific shaping, and material properties that will help meet future signature goals. Such structures will reduce radar observability, weight, corrosion, and maintenance requirements.

- New hull forms that are currently under investigation, some under Navy sponsorship, may include new wave-piercing hull forms with bow sections that look more like submarine hulls than traditional ship hulls. The combination of hull optimization and propulsion efficiency gains associated with electric drive may permit higher ship speeds if needed (perhaps 40 to 45 knots) and better seakeeping in rough water; this, in turn, will permit the use of smaller ships for a mission, so that ships can come closer to being sized for weapon system needs rather than to meet severe operational conditions, with resulting cost reductions.

Ship vulnerability to hits will always be a problem. Short of heavier armor or dynamic armor, the first of which will greatly increase ship weight and both of which will greatly increase ship cost, the best approaches are to reduce the chances for targeting a ship by signature reduction, and to reduce the chances that it will be hit, by active defense. Also, the chances of surviving a hit can be greatly improved by known design features such as separation of critical, redundant system elements (such as fiber-optic lines and instrument networks), and by automation in damage control that reduces response time and more accurately focuses damage control efforts. Finally, armor can be applied selectively to critical areas such as magazines and combat direction centers. All such measures are in train today in the design of modern warships, with serious attention to automation in damage control being the newest addition to the list.

Two main kinds of future surface combatant ship embodying these characteristics are visualized as promising candidates to meet emerging needs efficiently: a fleet combat ship that evolves from today's surface combatants, and a land-attack ship that evolves from the arsenal ship concept on which work is beginning. The two would have overlapping mission capabilities, with each specialized for different parts of the mission spectrum.

The fleet combat ship, eventually replacing today's guided-missile cruisers and destroyers, would be designed to engage in ASUW, ASW, AAW, and defense ATBM, and it could carry out power projection missions. It would emphasize sensors and defensive combat capability in the newly developed cooperative engagement capability (CEC) mode. Its weapon suite design would emphasize the sensors needed for these missions, including surface radars, sonars, and Aegis and beyond for ATBM. It would be able to operate helicopters for ASW and mine warfare, and UAVs for reconnaissance and targeting. It would have all the necessary links and nodes for C^4ISR within the overall naval force combat system. Its armament would include some 100 to 200 missile tubes, depending on ship size, loaded with weapons for ATBM, AAW, ASUW, and ASW; depending on circumstances of the time and on the missions assigned, ships of this kind might also be loaded with land-attack missiles, as today's fighting ships are. If missile technology advances meet expectations, such a ship may not need guns except for self-defense. Close-in defenses using laser weapons for defense against antiship missiles in the CEC mode, wherein the incoming targets can be illuminated from the side, may mature in time to be included in these ships' weapon suites.

The land-attack ship is visualized as an evolutionary advance from the arsenal ship concept. It would have 300 to 500 missile tubes loaded with missiles from the family of attack missiles described above, or a similar family that may evolve. The numbers of tubes will depend on resolution of questions about vulnerability and the advisability of concentrating too large an inventory of attack missiles in one platform. Since it will be an extremely attractive target, the ship may well need some close-in self-defense weapons, operated within the fleet in CEC mode so that extensive defensive targeting sensors would not be needed. The potential advantages may suggest building a land-target-oriented C^4ISR node (with input from external sensors) into the ship design to enable it to receive target information and launch missiles independently at times. This capability would enhance its flexibility as a combat ship oriented to prepare the battlefield for and to support operations of the land forces, and to operate in small surface combatant forces under some circumstances.

New Directions for Naval Force Aviation[10]

Air-delivered weapons will continue to be important in situations where

---

[10] Two types of aircraft are not treated in this discussion: armed helicopters, and maritime patrol

pilots on the spot are needed to perform functions—such as visual target identification in air defense or in armed reconnaissance, or response to the unexpected, as in close air support—where missile systems with humans elsewhere in the loop would not be responsive enough; where warhead weights needed are greater than ship-launched surface-to-surface missiles will be able to deliver; or where required depth of attack exceeds the sea-launched tactical missile range permitted under arms control treaty limitations. Aircraft (whether piloted or not) also have the advantage of being a reusable platform in situations that do not present an unacceptable risk of attrition, giving them an economic advantage for extended campaigns after antiaircraft defenses have been defeated.

Aviation is a major cost driver in naval force structure, warranting extensive attention to cost reduction both in acquiring aircraft and in the use of aviation in the combined arms context.

Advancing technology will offer many opportunities to improve aircraft performance while restraining cost increases or reducing costs. Microelectronic controls embedded in fixed-wing aircraft skins at flow transition zones will offer opportunities for boundary layer control that can increase lift, reduce drag, and consequently simplify high-lift devices like wing flaps. Increased turbine temperatures enabled by high-temperature metals will lead to higher thrust without increasing engine core diameters, or to smaller diameters for a given thrust. Both of these advances will permit expanding the flight performance envelope of future combat aircraft within a given gross weight and cost.

These advances will also permit lowering of takeoff and landing speeds to the 40- to 60-knot regime. Once that is achieved there will be a significant advantage in having airplanes take off loaded in short distances and land vertically after fuel and payload have been expended. With the increasingly high-thrust-to-weight-ratio engines that are expected, composite structures, and lightweight avionics, future aircraft designs may enable such performance with much reduced weight penalty. STOVL aircraft would not need to use the catapult and arresting gear. Thrust vectoring will help extend aircraft control into low-speed, high-angle-of-attack regimes not otherwise achievable, and will enhance combat maneuvering.

Stealth in aircraft design will always be needed for protection against proliferating air defenses. Especially, infrared signature reduction will be needed for

---

aircraft (MPA). Advances in both are expected to reflect progress that will be made in improving aircraft performance, "flyability," and maintenance, and, in the case of helicopters, stealth. The chief advances in armed helicopters will be reflected in the Army's Comanche program, which will likely define evolutionary directions of the Services' combat helicopter force for decades to come. At some point, it will be necessary to replace the P-3 MPA, which, although specialized for ASW, performs many other missions. The available, long-range transport aircraft that will exist when the need arises will form the basis of the new MPA, into which the necessary combat systems will be integrated.

protection of combat and logistic support aircraft against IR-guided shoulder-fired SAMs, and to protect aircraft against IR-guided air-to-air missiles. This problem will intensify as staring infrared arrays are incorporated into the weapons. Coatings to replace paint on aircraft skins and new nozzle designs will contribute to IR signature reduction at modest cost and little, if any, weight penalty.

Finally, advanced design and manufacturing techniques are expected to help in controlling costs as smaller numbers of aircraft are procured. These will include "electronic prototyping," derived from simulation-based design, to learn enough about designs to avoid costly changes after commitment to production; design for smaller production runs using expandable tooling rather than high-capacity tooling designed for high-rate production; and large unitary structures with composite materials having fewer parts and fasteners. These new approaches are being instituted in new aircraft programs today, and continuing progress can be expected under the pressure of resource constraints.

Combat operations using the new aircraft capabilities, and capitalizing on the presence of other technical advances and weapon alternatives for mutual support and expanding the mission spectrum, can be expected to influence how aircraft are used for combat. Defensive counter-air will, depending on circumstances, be able to take advantage of networked multistatic targeting techniques, enabling longer-range, cooperative engagements with air-to-air missiles and with surface-to-air missiles in the "forward pass" mode, in which aircraft or UAVs carrying the sensors pass target location information to the missiles. This capability to engage air threats at extended range would confer a great combat advantage on our air defenses, since U.S. and foreign short-range air-to-air missiles will continue to have comparable performance, detracting from any dogfighting advantage our superior aircraft would have. Positive identification will always be a problem. Developments currently being pursued in noncooperative identification will ultimately enable tracking of any airborne vehicle from takeoff to landing and maintaining a dynamic database of such tracks. UAVs with lightweight sensors will be able to observe other airborne vehicles and transmit what is seen in real time. These developments may, over the next 35 to 40 years, permit air target identification that is equivalent to visual identification by the weapon launcher without the need for visual contact.

Despite the great weight of fire that will be possible with the family of land-attack missiles, there will continue to be a need for close air support of troops in contact with the enemy. Close support aircraft, which may in the future be manned or unmanned, together with armed helicopters, on air alert or operating from forward arming and refueling points (FARPs) in the immediate rear of the ground forces, will be able to turn around rapidly and fly many sorties per day—on the order of 5 to 10—to greatly increase the weight of fire that can be brought against moving or dug-in opposition forces at critical points and times in an ongoing battle. Such surge capability will be needed to help sustain the rapid pace of future operations ashore. It would also have the advantage of using relatively inexpensive direct attack weapons in situations requiring great expenditure of munitions during a dynamic battle.

Fixed-wing aircraft able to perform this mission will have to move from their sea base on carriers and amphibious assault ships to shore with the forces they are supporting (as the AV-8B Harrier aircraft can do today). This means that they will continue to need vertical and short takeoff and landing (VSTOL) capability. If the reduction of weight penalty for STOVL can be achieved in future aircraft, the capability could be extended to combat aircraft for other missions, giving naval combat aviation great flexibility of operation from a variety of ships and land bases.

There will also be a mix of UAVs in fleet aviation. Some will be theater-level, high-altitude, long-endurance craft that will be needed in the fleet's vicinity for days on end. The UAVs may well be furnished by a joint agency, but they may be able to land and take off from carriers if carrier designs provide for such operation by aircraft with their very long wingspans. There would also be value in being able to refuel such craft from carrier-based tankers while they are airborne; this would turn them into a satellite analog, but one that is always available to the naval forces during an ongoing operation. Also, long-range UAVs, whether land-based or flying from carriers, may well be able to take over many of the missions of manned maritime patrol aircraft (MPA). Without people on board, their endurance could be extended indefinitely by the means described. Such a shift would mean revising the MPA processing system from on board the aircraft to one at a land base or on a carrier or other warship, and possibly melding some of the current MPA tasks with those of carrier-based support aircraft.

Additional UAVs will be developed for general targeting, airborne early warning (AEW), and providing communication relays over forward troops. The carriers will have to launch and recover such aircraft until ground operations move far enough inland to provide a secure rear area from which the ground forces can operate them. Finally, the ground forces are likely to have a family of combat UAVs to help in target location for close air support, in weapon control for "forward pass" weapon delivery, and perhaps for weapon delivery directly.[11]

Carrier design may change with the needs and opportunities to operate aircraft of the kinds described above. Carriers will continue to operate ASW aircraft. Manned AEW aircraft will be used for a long time before UAVs could

---

[11] The Air Force Scientific Advisory Board, in a recent study of future aviation technology (United States Air Force Scientific Advisory Board. 1995. *New World Vistas, Air and Space Power for the 21st Century, Aircraft and Propulsion Volume*, United States Air Force, Washington, D.C.) projected an uninhabited combat air vehicle (UCAV) design for weapon delivery in the mode of a fighter aircraft, in situations that are dangerous for manned aircraft. The Marines' combat UAVs might be of this character, or they might be of far simpler design; the implications for fleet aviation of having UAVs that launch weapons will be the same.

take over this mission, even if work on the UAVs were to start immediately. The need for general utility aircraft to bring cargo and personnel from shore to sea and return will continue. These aircraft may be versions of current and future ASW aircraft, or they may be derivatives of the Marines' V-22 tilt-rotor aircraft. There will also be value for the naval forces in acquiring a new-design heavy lift helicopter or functionally comparable vertical lift aircraft, tailored to carriage of containers as a replacement for the CH-53E when it reaches the end of its service life. The new helicopter would be tailored to handle logistic containers and the more rapid reloading at sea that containerization and other advances will bring. Carriers as well as amphibious support ships may also be called on to launch land forces in joint amphibious landings, as they were during the 1994 landing in Haiti.

Thus, carriers will become, even more than they are now, versatile, moving air bases at sea. Conceivably, if the STOVL combat aircraft can replace those in operation and being acquired today, if other manned aircraft functions such as ASW and cargo delivery all come to be carried out by vertical lift aircraft, and assuming that the UAVs can be designed to take off and land from a carrier deck in STOL mode (aided by the wind-over-deck derived from the ship's forward speed), it may be possible to design new carriers toward the end of the 40-year time period without the costly and operationally demanding catapults and arresting gear that help define carrier design today.

Finally, it must be emphasized that future design of carriers will be able to take advantage of all the technological advances in integrated ship instrumentation and automation, electric drive, and signature reduction that will characterize other surface ship and submarine design.[12] Thus, future carriers, including existing ships modified in periodic overhaul, will be able to reduce crew size and increase operating efficiency along with all the other ships of the fleet.

Future Submarine Design

Research and development has already provided reactor core lifetimes that eliminate the need for refueling during a submarine's service life—an important cost avoidance. Future advances in stealth, power density, and propulsion plant efficiency will be enabled by the development of electric drive and continuing research in nuclear plant design. Submarine design will benefit from the same advances expected in the design of surface ships, in distributed instrumentation, automation, and design integration that will allow crew reduction and more efficient use of the personnel on board ship. The advances in submarine capability induced by these and other technology advances will be impressive.

---

[12] The Naval Studies Board carrier study (Naval Studies Board. 1991. *Carrier-21: Future Aircraft Carrier Technology,* National Academy Press, Washington, D.C.) remains valid in describing in detail the potential application of these advances to carrier design.

The main determinant of future submarine design, however, will be the need to design the ships to execute routinely a broader spectrum of missions than have been assigned to submarines in the past. This will be reflected in aspects of the designs that affect the submarines' ability to carry out their missions while maintaining stealth, avoiding near-shore minefields, and maintaining communications with other forces. In addition to more shoreward orientation of submarines' mission spectrum, circumstances may arise in which opposing surveillance and defenses make it too dangerous for surface ships to approach closely enough to shore to provide sustained fire against inland targets and to carry out other power-projection missions. Submarines' stealth will, if they are appropriately configured, allow them to fulfill some of the vital power projection roles of the surface fleet, more safely and with less need for external protection.

Submarines will still undertake the traditional missions of ASW[13] and ASUW. They will also have to be designed as strike ships, able to launch any of the family of missiles described above. This will induce a significant design change moving well beyond the relatively few missile launch tubes in the bow of current attack submarines. Rather, the submarines are likely to be designed with payload sections comparable to (but easily distinguishable from) those of nuclear-powered ballistic missile submarines (SSBNs), including closely packed launch tubes and VLS technology adapted for underwater cold launch of the missiles for strike, fire support, or new missions such as ballistic missile defense.[14] In addition, the submarines will have to be designed to launch and recover UUVs routinely, and to launch or simply to control UAVs for various missions. UUV missions will include minefield reconnaissance, mine hunting and minefield neutralization, scouting for opposing submarines in ASW, offensive mining, intelligence collection and area surveys, and other tasks requiring underwater stealth. UAV missions will include targeting for the submarines' torpedoes and missiles, support of submarine-deployed special operations forces, and reconnaissance for information gathering in support of theater operations. The submarine system will have to be designed to maintain electronic,

---

[13] The ASW mission is discussed in the section below titled "New Approaches to Undersea Warfare."

[14] It may be argued whether, in the interest of preserving stealth and passive defense, submarines in the land-attack mission will simply launch deep-strike missiles against fixed targets and leave interdiction and naval surface fire support (NSFS) missions against moving or relocatable targets to surface ships, or whether, because the surface fleet may become too vulnerable in the early stages of a conflict, submarines will have to undertake the entire spectrum of land-attack missions. It can be similarly argued that surface ship vulnerability may favor the submarine as a forward-positioned missile launch platform for ballistic missile defense. Resolution of these arguments will have to await indications of threat development over the decades, and they may not be finally resolved until an active conflict presses the issue. For current purposes it is sufficient to note that over their designed service lifetimes future submarines may have to undertake those missions, so that the capability to perform them should be designed into the submarines from the start.

acoustic, or laser communication with these unmanned vehicles, regardless of the degree of autonomy that is built into the vehicles' operation away from the submarine.

Special operations forces fielded by the Marines or by the Special Operations Command will become more important in the coming modes of warfare described above and in the regional conflict study.[15] Such forces require stealth and support, and the size of units that may have to be launched and recovered by submarines will likely become larger than submarines have landed and recovered in past SOF operations. The capability to host and deploy these larger numbers of special forces will also have to be designed into future submarines.

Finally, submarines are and will continue to be ideally situated to gain information about actual or potential opponents using stealth to reach offshore observation positions while remaining themselves unobserved, and to engage in related information warfare activities. This will require sensor and communications systems related to those needed for the other tasks and missions described above, but augmented to meet additional needs imposed by the information-gathering and warfare missions.

Today's tactical submarines are able to carry out all of the above missions to some degree. Taken all together, with refinement and extension of the missions, the capabilities described above will lead to new multimission modularity in submarine designs that will significantly change their configurations and modes of operation.

Strike and Fire Support Evolution

The evolution of the surface fleet will depend on many economic and operational factors as well as the opportunities that technology will offer. The advantages of the family of missiles described in this section can be expected to encourage their proliferation as a weapon of choice for many naval force missions. This will affect the design of surface ships and submarines, it will influence how combat aviation is used by the fleet in strike, interdiction, and fire support, and it will influence how forces are configured to operate ashore.

For the naval forces to understand how these influences will act and to gain confidence in the new systems, they will have to implement the capabilities and use them in a variety of operations over a period of time. As indicated at the beginning of this section, this is, in fact, happening today. Also, the number of missile launch tubes in the Navy has been growing as new ships and submarines come on line, with the expectation that there will be about 7,000 tubes on about 70 ships just after the turn of the century, with more to follow as the planned 6

---

[15] Naval Studies Board. 1996. *The Navy and Marine Corps in Regional Conflict in the 21st Century,* National Academy Press, Washington, D.C.

arsenal ships are acquired. Consideration is currently being given to converting Trident SSBNs made available by strategic arms control reductions to a strike/SOF configuration; this would provide still more launch tubes that could be safely positioned near a hostile shore. There will be ample opportunity to load many of these tubes with newer versions of the land-attack missiles as they are developed, extending NTACMS and adding, for example, VLS-launched versions of the ERGM. As experience is gained and confidence grows in the planning for and utilization of these missiles in actual operations over periods of time, the resulting knowledge can be fed back into future plans to extend the missile family and adapt the forces suitably.

One of the criteria by which the value of a land-attack missile family, such as the one described, will have to be judged will be their overall impact on the naval forces' economic structure and the costs of carrying out major campaigns. Operational and technical differences among the systems make such comparisons difficult and dependent on many assumptions about scenarios, force maneuvers, targets, attack rates, weapon kill capabilities, and so forth. In addition to differences in tactical usage and effects, overall system costs would be key elements in the tradeoffs. The total costs of gun- and aircraft-based weapon delivery systems, with the costs of the munitions they deliver, must be compared with the overall delivery system costs together with the costs of the missiles themselves in the case of the missile-based systems. A detailed economic comparison among the systems was beyond the scope of this study. However, such an analysis, informed by the early operational experience described, will be essential for the Navy Department to ascertain the overall mix of weapon types that will maximize the naval forces' power projection capability within the budgets that will be available.

NEW APPROACHES TO UNDERSEA WARFARE

Antisubmarine Warfare

The marked reduction of U.S. research and development in ASW since the end of the Cold War has been paralleled by the increasing presence of two especially threatening aspects of potentially hostile submarine warfare:

• The continually improved quieting of Russian nuclear submarines and European-built diesel and air-independent propulsion (AIP) submarines, and the spread of these capabilities to other nations, some of which may become hostile; and

• Increased operation of U.S. surface Navy and logistic support ships in relatively shallow waters adjacent to potentially hostile coastal zones, in which ASW is especially difficult.

At the same time, it is likely that the marked reduction of submarine operations by nations of the former Soviet Union outside of Russian contiguous waters has led to a reduction in the training and readiness of U.S. ASW forces. The teamwork arising from the stimulus of the real-world experience gained in the interactions with those forces is a perishable capability that will have to be replaced in some other way.

Along with the evolving Operational Maneuver From the Sea concept that calls for logistic support of land operations from the sea with a much smaller or, in some cases, nonexistent land base, these factors raise the risk that an opponent could seriously interfere with a U.S. naval force expeditionary warfare campaign.

At about the time the Cold War ended, it was recognized that the conventional approaches to passive ASW were being negated by the quieting of Russian submarines, which had reached performance levels comparable to or exceeding the performance of U.S. nuclear attack submarines.[16] Modern conventional submarines submerged in deep water along coastal shelves are essentially undetectable by a single passive listener. Their noise output in the coastal environment is low, and reflections from the bottom and the surface and uncertain transmission paths make it very difficult to detect them at significant range even with active sonar.

Consequently, there was a move toward the use of low-frequency active (LFA) and explosive echo ranging (EER) ASW, and toward new designs of several kinds of deployable, distributed passive sensor arrays that, it was hoped, would allow the detection and tracking of the quieter submarines. Of course, the problem with monostatic active ASW is that in emitting a signal the emitter, which may be a submarine, reveals itself. Although EER systems mitigate this problem, their range thus far has been short, and so to be effective they require advanced information about where the target may be. Similarly, to be used efficiently, deployable arrays need cueing for placement and orientation so that they will be deployed in areas where there are submarines to be detected.

Future sensors (some of them MEMS-based), high-speed, high-capacity computing, precision navigation, and networking technologies will help in solving these problems. ASW is a cooperative enterprise involving a vast collection of different means to find and attack submarines:

• Surface combatant ships that have hull-mounted sonars, and also tow active and passive tactical sensor arrays;
• Ship-launched helicopters with dipping sonars;
• Fixed-wing aircraft, launched from carriers and from shore bases (MPA), that can drop fields of sonobuoys or floating acoustic sensor arrays to listen (or,

---

[16] "Text of Armed Services Panel Report on Naval Undersea Warfare R&D," *Inside the Navy— Special Report,* Inside Washington Publishers, Arlington, Virginia, March 20, 1989.

with an external active source, ping and listen) for submarines and then process the data on board;
- Long, densely populated sensor arrays towed behind especially configured ships (T-AGOS) for generalized surveillance of large ocean areas;
- Submarines with hull-mounted sonars and sensor arrays; and
- Fixed sensor arrays in large areas of the ocean, connected to processing stations at shore bases; and, to some extent
- Linkage among some of these sensors where possible.

Most of these means describe ASW capabilities built for passive listening to detect submarines by emitted noise. Nonacoustic means of detecting submarines successfully to varying degrees include cueing when the submarines leave their bases, magnetic detection, wake detection, detection of surface signatures created on passage through the water at depth, detection of emissions such as communications when they occur, detection of periscopes when in use, detection of snorkels of submarines that must breathe from the atmosphere when submerged, detection of surface-related activity such as the launching of weapons or landing parties, and detection of the submarines themselves from aircraft or spacecraft when they are near enough to the surface. The combat ships, helicopters, carrier-based fixed-wing aircraft, and MPA are also able to deliver antisubmarine torpedoes.

Submarine quieting degrades this vast array of capability to the point that the ASW force is capable of placing only small-diameter detection circles in the water, around sensors (fixed and mobile) that individually have only a very small detection range—perhaps as small as a mile or less, without the overlapping areas of coverage that would be needed for the sensors and subsystems to work cooperatively. In this environment the use of LFA ASW, together with increased emphasis on nonacoustic detection, is, of necessity, receiving increasing attention.

Future sensor, computing, and networking advances can contribute to alleviating some of the effects of quieting, alleviating the "alerting" disadvantages of low-frequency active sensing, and making cooperative use of the sensors more feasible. In addition, matched-field coherent signal processing that exploits signal amplitude and phase as well as variations in environmental conditions, made possible with future supercomputers, will permit extraction of much smaller signals from the ambient noise, thereby extending the range of passive detection. This type of signal processing is roughly analogous to the use of synthetic aperture radar (SAR) instead of simple monostatic pulsed radar, with similar improvements expected. Together with growing computer power, MEMS and other advanced sensor technology will permit very large and therefore highly sensitive and highly directional arrays, with tens of thousands of sensors connected by fiber-optic networks, to be built onto the sides of submarines. Similar techniques can be adapted to towed and fixed, bottom-mounted or moored arrays. Rough assessments brought to light during this study estimated potential increases of about 15 to 20 decibels in passive signal detection by these means, and the recovery of a significant fraction of the signal lost to recent advances in submarine quieting. Translation into increased detection range depends on specific ocean conditions, but the gains could be measured in miles under favorable propagation conditions. Exploitation of adaptive noise cancellation and beam formation should yield further improvements.

The same processing advances and computing power will enable multistatic active acoustic detection, using several sources on shore or on buoys deployed at sea. Improved processing will permit separation of interfering signals arising from multipath reflections and from reflections off false targets such as schools of fish. With appropriate coordination and timing, it will permit friendly submarines to position and align themselves to avoid reflections that would give them away. Detection ranges could be extended to distances on the order of 20 to 30 miles by use of multistatic active detection.

Finally, networking technology like that used in creating the cooperative engagement capability defense of the surface fleet will permit connecting all the sources of sensing and signal processing in a cooperative system that combines passive, active, and nonacoustic ASW. Like its electromagnetic counterpart that helps in detection of low-observable missiles and aircraft attacking the fleet and shore targets, a networked ASW cooperative engagement system will greatly advance the ability to find and attack hostile submarines beyond the capability of the individual means listed above.

Once a hostile submarine is found it must be attacked successfully. This outcome has been rendered more difficult with modern, quiet submarines (nuclear and nonnuclear) operating in the complex littoral environment and using sophisticated countermeasures. Advances are needed in antisubmarine weapons' sensors and guidance to improve detection of low-observable submarine targets, classify them against false contacts, cope with the highly variable acoustic environment, and overcome the countermeasures. Adversaries' submarines may also come to use the double-hull designs pioneered by Russian submarine builders. More powerful warheads are needed to attack such submarines within the same torpedo warhead package size as current air-delivered torpedoes. Meeting this requirement may be assisted by new warhead materials that will greatly increase the explosive power of torpedoes (and other undersea munitions). These same warhead materials will also be usable in mines and countermine munitions. Finally, advances in the undersea weapons that adversaries may use will require robust active and passive defenses for our ships and submarines, because the antisubmarine battle will not necessarily be purely one of hunter on our side versus hunted on the other.

Airborne nonacoustic detections will be fleeting and of relatively short range, but they will have the advantage of fixing the target submarine's position precisely. It will then be essential to be able to exploit this hard-to-come-by information rapidly, and rapid-reaction weapons must be developed for that purpose.

To be part of the cooperative ASW system, submarines will need to communicate with other fleet components. Meeting this requirement is easier now for them to do that than it was during the Cold War, since the risk of detection of antennas at or near the surface is lower, and the technology has advanced. Acoustic underwater communications are in R&D that can be used for short-range communications; the extraction of the communication signals from underwater reverberations and noise will be made possible by the same high-powered computing and processing techniques that will enable improved detection and tracking. Laser communications with submarines were also in work at the end of the Cold War and could be advanced for use in situations where water turbidity permits. Use of laser connections with distributed underwater communication buoys would also be possible if opposition becomes threatening enough. Radio communications using suitable antenna techniques will remain the means of choice for the submarine fleet to become part of the ASW cooperative engagement system, however.

History is replete with strategic disasters that resulted from failure to recognize emerging threats until it was too late to meet them. To avoid such an outcome arising from hostile transformation of what is viewed currently in many quarters as a quiescent submarine threat to naval force operations, it is important that R&D in the areas outlined above be continued at a level high enough to ensure successful implementation of ASW capability against the twin circumstances that have emerged to challenge our Cold War dominance of the field: submarine quieting, and operation in waters that are not conducive to the success of ASW methods used in the past.

Countermine Warfare

The other potential undersea expeditionary warfare "showstopper" for naval forces is mine warfare. All opponents trying to protect a shore against amphibious landings, or trying to deny free passage of warships and logistic ships through waters approaching their coasts, will use mines. Some of the mines will be highly sophisticated and hard to countermeasure; some could be deployed in a "smart minefield," in which diverse kinds of mines—bottom, floating, moored, or propelled and guided—might be controlled by a system of networked sensors that can trigger specific mines in a sequence that would inflict maximum damage on an approaching fleet or shipping train.

The Navy and Marine Corps have been well aware of this problem, and they have in work steps to meet it. The Chief of Naval Operations, in a December 1995 White Paper,[17] initiated a concerted Navy attack on the hostile mine war-

---

[17] Boorda, J.M., ADM, USN. 1995. "Mine Countermeasures—An Integral part of Our Strategy and Our Forces," White Paper, Office of the Chief of Naval Operations, Washington, D.C., December.

fare problem. This was to include integration of mine countermeasures forces with the fleet (instead of considering them as an adjunct to be called upon ad hoc); distribution of mine-hunting and neutralization capability among and on combat ships of the fleet; creation of the concept of mine warfare command ships, the first of which is operational; and increasing emphasis on mine warfare research and development.

Because the area of mine countermeasures (MCM) has been rather neglected until recently (in focusing on the Soviet threat, U.S. naval forces generally deferred to other NATO navies for MCM in forward areas), this study, as did the Naval Studies Board's 1992 study,[18] has focused on the use of available assets and technology to create a major capability to deal with the area. The capability thus derived will take time to build, but once available it should serve our naval forces well for an extended period.

First and foremost, attention is needed to ensure availability of intelligence, surveillance, and reconnaissance: intelligence to know in detail what mine warfare capability any operation will face; surveillance using all available assets to track mining activity and to gain the options of mine interdiction and mine avoidance; and reconnaissance to provide ground truth confirming unmined areas or to concentrate MCM forces only on areas known from both surveillance and reconnaissance to be mined. There are now Navy and joint programs, which must be supported, that aim at providing this capability.

To allow a battle force to proceed independently, an organic MCM capability, resident on combatants and support ships, must be in place. Assignment of a force able to deploy MCM-capable, or adaptable, helicopters with the ability to carry and use modular mine-hunting and mine neutralization equipment, remote mine-hunting undersea vehicles, possible new developments for mine neutralization such as acoustic pulse power if it becomes successful, and the offshore and surf-zone clearance capabilities that are described next, would provide a battle-force with the needed countermine protection. In addition, attention must be given to passive countermine measures, including a serious, steady program to reduce and control (or eliminate, where possible) the magnetic and acoustic signatures of ships, and attention to reasonable hardening of ships to the effects of mine detonations.

For mine clearance to the surf zone, surface craft of up to 30 tons' displacement can perform the mine-hunting, mine neutralization, and mine-sweeping functions. In the past such surface craft have had limited speed and range and have been limited by sea state; however, the SWATH hull form offers a solution to those problems, permitting operations in sea state 3 or 4 and including survival in higher seas. A SWATH MCM platform of up to 30 tons in size could be designed to be transported, launched, supported, and recovered, along with air-

---

[18] Naval Studies Board. 1992-1993. *Mine Countermeasures Technology*, Vol. I-IV, National Academy Press, Washington, D.C.

borne MCM (AMCM) helicopters, by a ship similar in size and general design to the LSD-41. It is expected that a ship with such general characteristics could accommodate 10 MCM surface craft and 10 fully hangered AMCM helicopters. Thus, one ship capable of deploying with an amphibious ready group (ARG) or battle group can bring to bear the MCM capability of roughly 20 MCM-1 or MHC-51 ships. Additionally, this MCM-carrying ship could act in the capacity of a command ship for MCM operations and be fitted with appropriate communications, analysis, and command capabilities. Finally, an expendable mine neutralization vehicle (EMNV) can reduce the classification-to-neutralization cycle time, by precisely placing charges to achieve sympathetic detonation of the mine's main charge, and to complement the capability of small mine hunters and AMCM helicopters.

The most difficult of all mine scenarios is in the surf zone (SZ) and craft landing zone (CLZ) that amphibious landings must transit. This area can contain a high density of diverse mines mixed with an equally difficult array of obstacles, while speed and flexibility of clearance are mandatory. The rocket-propelled line charge (SABRE) and explosive net (DET) being developed by the Navy and Marine Corps will find use on beaches having no obstacles, and in neutralizing minefields on land. Additional "brute force" methods would greatly strengthen the naval forces' capability for rapidly clearing the SZ and CLZ immediately in the path of an amphibious landing, and shortly before the landing. One, which has been described in prior Naval Studies Board studies,[19] would offer the only means for almost instantaneously clearing mines and obstacles from a 50-yard-wide channel to and across the beach. This would use large (e.g., 10,000-lb) precision-guided bombs dropped in a line set in GPS coordinates and exploded simultaneously. It would not necessarily explode all the mines blocking the channel, but it would at the least throw them and any emplaced obstacles to movement aside and set up a pair of berms that landing craft could use, with GPS assist, to guide themselves through the channel. Calculations, modeling, and limited field tests since 1992 have tended to confirm original estimates that all mines and obstacles can be excavated from a 50-yard channel by this method, known as Harvest Hammer. Smaller bombs might be used, requiring more sorties by combat aircraft; the critical elements of the technique are the accurate placement and timing sequence of the explosives.[20]

---

[19] Naval Studies Board, 1992-1993, *Mine Countermeasures Technology,* Vol. I-IV, National Academy Press, Washington, D.C.; and Naval Studies Board, 1996, *The Navy and Marine Corps in Regional Conflict in the 21st Century,* National Academy Press, Washington, D.C., p. 85.

[20] Recent Service analyses of similar approaches have been discouraging in terms of the number of aircraft sorties required, but they did not examine the problem in the terms described here. A test in the United Kingdom using emplanted charges has given encouraging results. Use of the heavy bombs would obviously require use of the USAF bomber force to deliver them in a joint support mode for amphibious landing, although if projected improvements in the energy of insensitive explosives are achieved, then organic aviation would be able to perform the mission.

Mine countermeasures is the only warfare area that operates in daylight hours only (with a limited exception in the Persian Gulf). The AMCM helicopters cannot operate at night because they lack artificial horizons and night vision equipment, and the surface ships do not operate out of concern for floating mines. Installation of appropriate night operating equipment on the AMCM helicopters, and floating mine surveillance and neutralization provided by airborne light detection and ranging (LIDAR) and the Rapid Airborne Mine Clearance System (RAMICS) supercavitating projectile, could double, or even further improve, the effectiveness of the available MCM forces.

Another approach would attack many mines in parallel, rather than hunting for mines and marking those found for later destruction, one at a time. An early proposed implementation of such a scheme is embodied in the Defense Advanced Research Projects Agency's (DARPA's) "lemmings" concept, in which a mass of small crawling vehicles disperses over the ocean bottom, each one recognizing a mine it may encounter and then detonating at an appropriate time to destroy the mine. As currently articulated, in some scenarios the effectiveness of the lemmings concept may be reduced by countermeasures such as underwater fences, but the concept opens an R&D avenue that holds promise of more rapid mine field neutralization over the coming years and decades. A related approach, perhaps using the same principle of parallel attack, may be possible using coordinated UUVs, probably operating with a degree of autonomy but under general ship, submarine, or aircraft control.

High-pulsed-power techniques have been proposed to destroy mines from a distance. Power pulses from a single source would have to be very large and very close to the mines to be effective, but techniques to focus the power from several lower-powered pulses have been proposed and are currently entering exploration. It will be some time before it is known whether the techniques can be made to work in a disturbed aquatic environment.

The key point in the entire countermine warfare area is to recognize that mines are likely to defeat expeditionary force plans at critical times, and that avoiding that outcome with high certainty requires appropriately funded R&D focused on a large variety of methods, including those newly proposed as well as the older ones already in work, and accorded sufficiently high priority.

NEW APPROACHES TO OPERATIONS IN POPULATED AREAS

Armed conflict along the littoral will frequently take place in populated areas, control of which is often one of the main objectives of military action. Such operations may vary from evacuation and rescue missions to the capture of a city to use its port and airfield facilities and to prevail over the governing apparatus of a country. The opposition may vary from a small band of terrorists to regular army divisions.

Typically, once assault rather than siege becomes the tactic of choice, capture of populated areas can entail many friendly casualties, many casualties among the resident civilian population, and much incidental destruction. Every substantial building in a heavily populated area can be turned into a small fortress; if it is reduced to rubble, the rubble favors the defense. Sewers, fences, and irregular street plans or winding suburban roadways, often lined with thick vegetation, afford defensive cover. Taking a populated, built-up area without causing heavy civilian casualties may mean fighting with small arms from street to street, building to building, and room to room and is certain to result in high friendly casualties. Such fighting characterized World War II; it was experienced in driving the Viet Cong out of Saigon after the Tet attack in 1968; it was seen again in the Russian attempt to capture Grozhny, in Chechnya, in 1995. It has been a universal characteristic of 20th-century warfare in populated areas.

Denying War-supporting Capability

Modern and future technology will offer many means to avoid the worst of these characteristics of military operations in populated areas. The nature of the attack may determine the means used. If it is desired simply to greatly reduce the ability of a heavily populated area to support a war effort, this can be done by precision attack against the facilities that support the area: its power stations; its major transportation nodes (bridges, tunnels, rail, and aviation control points); and its communication nodes. Such attacks, which can be made by appropriately armed land-attack missiles if not by aircraft with the proper weapons, need not destroy the facilities completely; they need only incapacitate them severely by attacking their most exposed and vulnerable elements. In addition, the target area will be vulnerable to information warfare using diverse media, to confuse the leadership and to render their popular support ineffectual. Even urban areas in primitive countries will have such vulnerabilities and will not be able to function effectively to support their populations, much less to support national war efforts, if such critical targets are taken out of action.

Knowing the Local Area

If a major populated area or a part of it must be captured or secured, emerging and future technologies will permit doing so with far fewer casualties and less destruction than has been seen in the past. An essential prerequisite, however, is extensive and accurate local intelligence and an understanding of the culture in the local area by the entering forces and their leadership.

Without local knowledge, attacking or occupying forces are likely to be subject to unexpected and deceptive tactics, sneak attacks and unexpectedly effective defense, and the confounding effects of hostile civilian actions. The local knowledge required involves more than knowing the layouts of streets, facilities, and buildings, although those are required, even to specific construction details of key buildings. It is also essential to understand the local culture, in order to understand what local tactics and doctrines may arise from local history, to anticipate how local forces may manipulate or hide within the civilian population, and to understand the kinds of psychological operations, appeals, or threats—through the media and otherwise—the local population will respond to, and how they may respond. Ideally, the backgrounds of local leaders will be known, so that it is understood how they operate and how they may be thwarted before they attempt hostile counterattacks. Knowledge of the local language will be extremely valuable, but by itself should not be taken as a substitute for deep local knowledge. None of this is different from the knowledge needed for successful warfare anywhere; it is rendered especially important by the stakes, in casualties and length of war, that are involved in military operations in populated areas.

Building this kind of background will require local expertise. There will be no substitute for effective intelligence, informed by area expertise derived from trusted sources that have been proven reliable. This expertise may often be found within local or coalition forces, but must then be treated cautiously lest local political objectives distort the knowledge transmitted. Local intelligence networks that can be called into play when needed will make invaluable contributions. All this may take more time and advanced preparation than the development of a particular crisis or action will permit. Planners will have to anticipate where such actions may take place and start early to build long-lead-time elements of local knowledge. Although intelligence resources may be limited overall, the cost for building area expertise, even if some of the effort pertains to areas where it is ultimately not needed, is small relative to the payoff for having it or to the loss incurred if it is not available when it is needed. The task must be joint, because joint forces will inevitably be involved, so that the naval forces will not have to absorb the expenses all on their own. The Department of the Navy must take the lead in initiating the joint intelligence preparation for expeditionary warfare contingencies along the littoral, however, since they are likely to be the first to need it on the spot.

Tactics, Weapons, and Techniques

Once the necessary local knowledge is available, the capture of populated areas in the future will depend on our forces "operating smart" with advanced technical means, rather than using massive force. In this approach, major forces would surround the area to be taken, to blockade it and to be positioned for later entry to secure it, but they would not enter against potential opposition.

Small, platoon-sized units would penetrate early in conjunction with information warfare and psychological operations to neutralize defending forces, using their area expertise and intelligence with helicopter and light armored vehicle mobility (or boat mobility in areas with waterways) for decisive positioning of forces, rather than for direct attack. They will be able to use advanced sensors, including covertly distributed MEMS-based unattended sensors with GPS location and broadcast capability, building-penetrating radar, acoustic sensors, and infrared scanners, to locate opposing forces as precisely as possible before attacking them—to the point of knowing which rooms in a building are occupied. They will be aided by extensive use of robotics, such as small, sensor-carrying remotely piloted air vehicles of various sizes and unmanned ground vehicles, for scouting blind streets and other areas, for denying pathways, for decoying, and for placing explosives or otherwise attacking and destroying targets. They will be able to operate mainly at night, when even if opponents have night vision devices it is easier for an attacker to sow confusion and create disorganization.

Among the weapons being devised for the attacking forces in the future will be extensive nonlethal or less-than-lethal weaponry to incapacitate rather than to kill or wound opponents. Those within the realm of possibility include means to render people dysfunctional individually or in groups, through activation of such means as disabling sound levels, nausea-creating agents, and sticky, slippery, and wetting substances, and by rapidly erecting barriers to movement in the form of helicopter-emplaceable quick-hardening foams or other rapidly emplaceable barriers.

Many of the above means to neutralize defenders of populated areas may not work against large and heavily defended cities—capital cities defended by hostile and well-armed divisions that are not loath to use armor and artillery, for example. But even in those situations, the means described may be used to take a city by sections from the outside in if the time is available and there is value in doing so. The means described for disabling a population center can help to shorten the time by weakening resolve to resist. More to the point, however, those situations represent one end of a continuum that has rescue of hostages and defeat of terrorists holding specific facilities at the other end, and many stages of military action in actively or potentially hostile populated areas in between. No one would argue against preparation to deal with most of the spectrum because one end of it may be especially difficult when using the means described.

The naval forces will need all of these advanced information and technical capabilities, ranging from means of disabling infrastructure and obtaining deep local knowledge to ways of capturing hostile areas with minimal friendly and local casualties, as an essential part of their "kit of tools" for expeditionary warfare and operations other than war.

REENGINEERING THE LOGISTIC SYSTEM

Logistics is usually considered as an "annex" to military operational plans. However, logistic considerations determine what operations can be undertaken, when they are undertaken, and the extent to which they will succeed.

The emerging Navy and Marine Corps concept for Operational Maneuver

From the Sea in expeditionary warfare incorporates a radical change in logistic concepts. Instead of building a massive logistics base ashore to support subsequent ground-force operation, with attendant time delays and protection requirements, most of the logistics base is to be kept at sea, at least for the early stages of any operation. When it is moved ashore it may not be as massive as logistics bases have been in the past—it may contain enough supplies for a week or so, rather than 60 days' worth. Logistic support for the combat forces is to be provided on an "as needed" basis, with a base ashore to support surges according to operational need. In the new logistic system, there will be far less redundant supply. This feature parallels and relates to changing concepts of logistics and support in the civilian economy that are being driven by economics, advancing transport, manufacturing, and system management philosophy, and associated technological developments in the information and transportation areas.[21] These changes will also be reflected throughout the joint logistic system for supporting forces in theater.

The changes in the logistic system that are called for will demand more than marginal improvements achievable through occasional renewal of system elements like shipping. They will require changes in logistic concept, priority, and equipment at all levels, and a long-term strategic plan for achieving the changes in parallel with the changes in the combat forces that are to be supported.

Achieving efficient logistics will depend in part on reducing the logistics load. This means incorporating distributed, computer-assisted advances in system readiness, maintenance capability, and support based on extensive and readily available information about the status of systems and supplies, and taking other steps to reduce the total amount of supply to be delivered. It will also entail significant changes in system design for loading, moving, and delivering essential support for forward combat forces. To assess the nature of these changes in the logistic system, it is necessary to assume certain factors as givens:

- Operational Maneuver From the Sea, in some evolved form, will become the standard naval force expeditionary warfare doctrine.
- Maritime prepositioning forces (MPFs) will continue to be used into the indefinite future.
- Intercontinental and local force logistics will both be fully integrated into the worldwide information system and communication network with message priority equal to that of tactical communications. (Logistic communications to and from forward forces in a "supply as needed" combat situation are tactical communications, not the pipeline-filling transmissions that have characterized logistic communications loads in the past.)

---

[21] For a detailed discussion of the impact of OMFTS on naval force expeditionary warfare logistics, see Naval Studies Board, 1996, *The Navy and Marine Corps in Regional Conflict in the 21st Century,* National Academy Press, Washington, D.C., pp. 69-81.

From this base, logistic support of the naval forces may be considered in terms of readiness, support of forces at sea, and support of forces ashore.

Information-based Readiness and Its Impact on Logistics

The term "information-based readiness" has been coined to describe a logistic system that will capitalize on the use of computer-based design and management in all activities that create and support major military systems, and that make and keep them ready for operation and combat. This concept constitutes a major application of the enterprise process technologies discussed earlier. It implies concurrent incorporation of logistic support with operations as part of an end-to-end simulation-based system design process for all military systems. Information-based readiness will then require sensor-monitored performance of all weapon system platforms and stored weapons such as missiles, for condition-based, rather than schedule-based, maintenance. Parts will be supplied as needed, and some may be manufactured in forward areas by agile manufacturing techniques. This approach will mean significant changes in the transportation systems to ensure ad hoc movement from points of origin to supply nodes and subsequent delivery of diverse goods directly to using forces, rather than routine, a priori bulk delivery to a central storage point. The new capability could not be implemented without modern computing power. For forward forces anywhere in the world, there will be computer-based, distributed training of repair personnel, computer-based troubleshooting, and distributed troubleshooting expertise available on call from system design and integration contractors or the few rear-area military support depots.

Supporting Forces at Sea

The main loads that must be delivered to ships at sea are fuel and ammunition. Two approaches to easing the resupply problem for fuel are to reduce the need for fuel and to move the fuel that is needed more efficiently.

Reduction of fuel use at sea would reduce the frequency of refueling, which takes ships out of action for significant periods of time. The need for ships' fuel will gradually be reduced by incorporation of more efficient electric drive and hull drag reduction in the major platforms. Aviation fuel needs will gradually be reduced as engine efficiencies, reflected in reduced specific fuel consumption, increase in new and upgraded aircraft. Such changes may not be reflected in a markedly reduced need for fuel resupply in wartime, when all systems are pressed to the limits of performance. However, even small improvements in efficiency will mean significant cost savings over system lifetimes, and they could enable extended operations under some wartime conditions when even modest increases in time between refuelings could confer a tactical or operational advantage. There may also be reductions in aviation fuel use as the mix of naval force aviation changes along with changing combat techniques and systems; this will be difficult to predict, and the differences will have to be assessed and reflected in changes to the logistic system as the forces evolve.

Ammunition resupply requirements will also change as the means of land attack and fire support change. Shifting strike and fire support from "dumb" bombs and shells to greater use of guided weaponry, and using large numbers of tube-launched weapons for strike and naval surface fire support, will radically affect those requirements in currently unpredictable ways, leading to many changes in logistic support loads and how they are delivered. A major operational problem, capable of technical solutions but needing system analysis of the design and operational tradeoffs, will be whether to reload missiles into ship VLS at sea, or to return the ships to the nearest base for that purpose after all or parts of their loads are expended. Exploration of this problem must become part of the overall, simulation-based system design for the surface and undersea land-attack ships and forces, possibly arriving at different solutions for each type of force.

For the remainder of the logistic load, reduced crew sizes will be reflected in a reduced logistic train from CONUS to the fleet. More efficient and rapid delivery to the under-way replenishment ships can be achieved with containerized loads, saving at-sea manpower and preparation time. With a move to containerized logistics, the next generation of logistic ships will have to be designed so that the loads can be broken out for "retail" delivery to diverse warships at sea. This is consistent with the changes needed for OMFTS.[22] Faster fuel-pumping capacity that is in development will also reduce the time spent in refueling, rearming, and resupply operations.

Solid-waste management has also become a major problem for ships at sea, as constraints against ocean dumping of such waste increase. System-based solutions will be required in new ship design: designing for reduced waste in the first place, and consideration of on-board treatment, compacting, and storage for shore disposal, incineration, or a combination of these methods.[23] This will have to be considered part of the overall logistic system in designing for support of ships and aircraft at sea.

Supporting Forces Ashore

As for ships at sea, the greatest loads to be moved ashore during combat operations are fuel and ammunition. In most environments, uncontaminated water also represents a significant load, difficult to process in mobile operations

---

[22] Naval Studies Board. 1996. *The Navy and Marine Corps in Regional Conflict in the 21st Century,* National Academy Press, Washington, D.C., pp. 74-75.

[23] Naval Studies Board. 1996. *Shipboard Pollution Control: U.S. Navy Compliance With MARPOL Annex V,* National Academy Press, Washington, D.C.

and difficult to deliver from outside. Fuel needs ashore will be mitigated to some extent by the use of more efficient power sources for support equipment—long-life batteries and fuel cells instead of electric generators, for example—although the fuel needs are associated mainly with combat vehicles and aircraft.

Although fuel transport by pipeline from ships or depots will be available to support forward ground forces in stable situations, combat operations will be wide ranging and will require air resupply, or resupply by tanker trucks when stable and secure land lines of communication are established. Air resupply of fuel and water to mobile forces during combat will depend heavily on the use of 500-gallon pods slung under heavy-lift helicopters and V-22 aircraft, both of which will be able to carry more than one pod per load. This (and other air resupply) will require protection of the air routes of supply, and landing zones—forward arming and refueling points (FARPs) and forward troop positions.

A reduction of massed artillery fire in favor of fire support from the sea, as visualized in the OMFTS doctrine, will significantly reduce the daily ammunition load that must be delivered. For example, the regional conflict study estimated that the logistic load to support a light battalion-sized force ashore would be reduced from 37 to 7 tons per day if all the battalion's fire support were delivered from the sea.[24] Land combat units with less heavy equipment, as visualized under the evolving doctrine, will also require less fuel.

Remaining logistic requirements for the ground forces in combat will have to be supplied routinely (for food and other consumables) or ad hoc (for maintenance items), usually by air. Air delivery will involve vertical-lift aircraft, with the same protection problems posed by fuel and ammunition delivery, and sometimes precision air drop using systems that are being developed by the Army and Air Force.

In addition to reducing the loads as described above, the ground forces will have to practice "smart" logistics to ensure steady resupply as needed with minimal waste in the system. This will, as will ship resupply at sea, require containerization starting from the sources in CONUS, and continuous visibility into container contents through electronic tagging and tracking until delivery to the using units. Logistic and MPF ships will have to be designed with the capability for on-board container handling and load manipulation, and for operating vertical-lift aircraft. A heavy-lift helicopter to replace the CH-53E, when that is needed, should be designed to move containers from ships to operational forces ashore, less awkwardly than in the current process for large underslung loads. Not least, integral container carriage will allow such aircraft to fly closer to the terrain in areas where very low altitude flight is needed to afford a measure of protection from shoulder-fired SAMs.

There will be times when logistics delivery over-the-shore (LOTS) will be

---

[24] Naval Studies Board. 1996. *The Navy and Marine Corps in Regional Conflict in the 21st Century,* National Academy Press, Washington, D.C., p. 70.

required. Currently, such delivery is limited to relatively calm seas—sea state 2, or waves of 3-ft height or less—permitting over-the-shore offloading only about half of the time in areas around the world where military operations are likely. There are means in work or proposed for increasing LOTS capability for offloading through sea state 3, or waves up to 5-ft height. This advance would increase the period of time in which offloading could be conducted over the shore by 20 percent or more in many areas, depending on the geographical area and the season. The means in work include stable cranes, high-sea-state lighterage, and "portable ports" or emplaceable causeways that will permit docking and offloading of combat vehicles and load transporters.

Many aspects of the logistic advances described can be implemented using existing technology. In areas such as containerization and container handling, loading and unloading at terminals, and asset tracking, the commercial world is ahead of the military. The latter can adopt and adapt the technology applications it will need. In doing so, it will have to ensure that compatibility is retained between the military and commercial systems, in case the latter must be called on to augment the military logistic system—much in the manner in which the Civil Reserve Airlift Fleet (CRAF) is used.

Logistics and support, in addition to communications, are areas where extensive "outsourcing" and privatization will take place, in the interest of conserving resources and improving efficiency. This will add to the use of COTS systems and technology that will be adapted to many military systems. All of this trend reinforces the argument for extensive efforts to ensure functional and physical compatibility between military and commercial systems.

MODELING AND SIMULATION AS A FOUNDATION TECHNOLOGY

Over the years since World War II, mathematics and computer models have been used increasingly to describe the dynamics of military engagements and warfare. Simulated equipment and computers have enabled representations of military equipment and operations. Modeling and simulation (M&S) now constitutes a fundamental technology area underlying all aspects of the creation and use of military systems and forces. Three basic kinds of simulation that are used by the military forces reinforce and interact with each other: (1) so-called constructive simulation of systems and combat performed wholly on computers; (2) distributed interactive simulation (DIS) and "virtual" simulation that join actual or simulated equipment operated by people—many of them in different locations and networked together—with computer-generated "environments" to simulate operations of the systems and their use in the field; and (3) simulations of combat (field exercises) in which military units with their actual equipment operate in the field on instrumented ranges, with quantitative measurement of system and unit performance. All DIS, virtual simulations, and field exercises have people in the loop by design, and constructive simulations have also been devised to involve people for decision making.

The various techniques involved have also been developed by industry to support design and construction of military as well as commercial systems. Applications vary from exploration of preferred system design parameters to simulation, derived from computer-aided design practices, of system elements or complete platforms—aircraft, ships, or manufacturing plants—to examine how internal space is utilized and how the systems will perform under various conditions.

All these forms of simulation are now used in complex combinations. They affect all aspects of naval force planning, acquisition, and operation: designing systems and optimizing their operation; choosing among systems and forces for specific military tasks; developing and testing operational concepts with real or postulated force designs; mission planning and rehearsal, and evaluating alternative courses of action in carrying out missions; evaluating mission outcomes and the results of operational test and evaluation; and training forces and commanders at all command levels.

Such a pervasive technology requires a new "corporate" management approach if the naval forces are to capitalize fully on the benefits that modeling and simulation can offer. These include the ability to evaluate and to integrate ideas, systems, and force designs and to adjust them to each other before actual building begins, as well as to evaluate the economies to be gained by eliminating steps in building and modifying hardware early in the creation of military systems and forces. As was the case in prior years for the technology of computing itself as it was being integrated into commerce, industry, and the military forces, it is now becoming apparent that M&S demands the attention and support of top Department of the Navy command and management levels because it affects every aspect of military force design, equipment, and operation. The necessary integration of viewpoint and utilization cannot "just happen" without such attention and support.

The Joint Chiefs of Staff have recognized this in arranging for the construction of large-scale simulation models—JWARS, to support the requirements and process of force design, and JSIMS, to support education and training and their integration into military operations. The Navy and the Marine Corps have been building their own separate management and operational structures for M&S and establishing simulation systems for the individual Services. The latter include, in addition to the use of M&S in weapon system design, the Navy's Battle Force Tactical Trainer (BFTT), a simulation of maritime operations (MARSIM), the Naval Simulation System (NSS), and the Marine Corps Commandant's Battle Laboratory that will, among other things, systematically test and help develop the evolving OMFTS concept. The Navy's cooperative engagement capability was developed using "embedded simulation" by operation of actual air defense systems aboard ships at sea and defenses on land against simulated attackers.

Completely incorporating and effectively using M&S as a Navy Department foundation technology requires the creation of a joint Navy and Marine Corps strategy that spans the two Services' operations in expeditionary warfare, where the two must function as a single force that operates in a joint environment with other Service forces involved. This strategy and the M&S activities it guides and supports must also feed, draw from, and interoperate with the joint efforts embodied in JWARS and JSIMS.

After completion of the institutional arrangements by which the Department of the Navy can best capitalize on M&S, two important advances (which could be undertaken simultaneously) are needed: (1) bringing the M&S conceptual foundation up to date with current knowledge of how modern warfare is and may be fought, and (2) changing the technical basis of M&S to incorporate and capitalize on modern computing and M&S technology. The needs for these advances apply initially in the area of constructive simulation but also will have an important influence on the way virtual simulations and field exercises are planned and on the way their results are interpreted and used. There is at present a dearth of theoretical understanding and knowledge of modern, post-Cold War types of warfare based on collected and analyzed data to describe the phenomena of warfare—what really happens in complex interactions among modern armed forces and between them and irregulars of various derivations, why it happens, and what drives the effects of the critical parameters. Indeed, the databases on which such a theoretical foundation can be built have yet to be assembled.

As a result, while computer programming and software technology have advanced rapidly and have been used to build today's generation of models and simulations, the knowledge base on which the existing models and simulations are built is obsolete and deficient in many ways. For example, many models derived from years of development still do not allow for dynamic evolution of a battlefield or a battle area and feedback into force operations, and their output in the hands of users not familiar with their multitudinous and usually hidden assumptions often does not accord with modern understanding of force-on-force interaction. As another example, simulations that attempt to describe the functioning of individual systems or subsystems in exquisite detail both challenge the economics of efficient computing and miss the mark in simulating the functioning of networked systems with many similar components that can each be described by a few functional attributes.

Decision makers who rely on M&S for system acquisition or military planning have little basis, at present, for knowing whether the M&S results that they use are valid representations of the real world on which to base extrapolations to some future world. There is a dearth of model validation that compares the results of models describing warfare with the outcomes of actual conflicts or even of field exercises, nor are there credible methods for model validation. If there is to be any confidence in the projections and plans that the M&S results are supposed to support, there must be a continuing effort to validate existing and new models against real-world situations when there are data for comparison, however sketchy or anecdotal. Building databases that include historical data from actual warfare and from pertinent exercises will be an essential part of such an effort.

Recent simulation concepts being developed in the commercial world, and the growing mass of results from virtual simulations and measured field exercises in the military world, can help rectify some of these deficiencies in modeling and simulation. To make the most of the new concepts and data, much of the current approach to and utilization of models in planning will have to be changed, often at the price of extensive investment in replacements for current models, simulations, and M&S tools. Future practice should create an interlinked, hierarchical family of models, all developed together, describing various levels and phases of expeditionary warfare from the system through the strategic level. Such a family of models would be based on a common high-level architecture and a common set of input data. The various models would be calibrated together and have functional connections to allow various elements of the family to operate together in diverse combinations.

Use of M&S in the military environment where there will be great uncertainty about opposing forces and operational environments far into the future must allow for that uncertainty. Within the family of models and simulations, it will be necessary to provide the capability for easy and inexpensive exploratory analyses and tests with different scenarios, databases, and concepts of operation, to learn which approaches are most likely to give robust solutions before specific plans and force designs are "cast in concrete."

These advances in the M&S field to support naval forces will not be made effectively without focused technical support. As in any other important technical area, an ongoing research effort is needed to provide that support. This research must first be focused on military science and technique, to ensure that the knowledge base incorporating the latest concepts and understanding about the uses of naval forces and how they will fight is included in the resulting models and simulations. Research must be performed in simulation science and technology applicable to military systems and operations. And databases covering worldwide military forces and environments, by warfare area, must be constructed and maintained. This research program would, especially, review and resolve technical problems in adopting and adapting related developments from civilian areas that can be applied in military M&S.

FOCUSED RESEARCH AND DEVELOPMENT

No modern, technology-intensive enterprise can prosper without sustained research and development support focused on the enterprise's main objectives. This truism has been recognized for the armed forces since World War II, but the nation may be losing sight of it today as budget concerns move front and center in national attention. The environment in which future naval forces will exist and in which they will have to function effectively will be characterized by continuing budget stringency, barring the emergence of some future mortal threat to the United States and its allies. Regardless of the level of resources that will be allocated to support the creation of the entering wedges of capability that this study foresees as essential to future naval force viability, and however they are found, the R&D part of those resources will have to be spent as efficiently and effectively as possible, and in a timely manner.

In addition to effective technical management, a key step in effective use of resources for R&D will be to focus the R&D effort on those elements that are unique to military and naval forces, and for the rest to capitalize to the greatest extent possible on R&D and technology emerging from the civilian, commercial sector. The technology areas listed in Table 6.1 were reviewed to see where relevant R&D is currently performed and is likely to continue. The review showed extensive scientific and technology development effort in the civilian sector that can be of value and use to the naval forces in the following technology areas or clusters:

- Information technology (with some exceptions to be noted),
- Technologies for human performance,
- Computational technologies,
- Automation,
- Materials (with some exceptions to be noted),
- Power and propulsion technologies,
- Environmental technologies, and
- Technologies for enterprise processes.

Although particular areas of science and related military and naval applications will always require Department of the Navy investment and attention, military R&D in the above areas can concentrate heavily on adapting the civilian and commercial technologies and their products to naval force use.

This orientation must be adopted with caution, however, because in many areas commercial industry is also deferring long-term R&D in favor of short-term programs offering a quick payoff in highly competitive markets. The Department of the Navy must thus remain vigilant to ensure that its needs will indeed be met in these areas by the civilian world. In no sense, therefore, should comments on priority in this regard be taken as a suggestion that basic, long-term research be foregone by the Department of the Navy in all these areas *without first ascertaining that research needed for naval force purposes will in fact be performed by the commercial sector.* The Navy Department must also be ready to recognize and adapt wholly new advances that can change how military tasks are performed, equipment is brought into being, and kinds of equipment created. The naval forces must remain open to new and vital knowledge. The issue is to apply appropriate judgment to allocation of scarce research resources.

With due attention to these caveats, it appears now that science and technology for military and naval force use will have to be especially sustained by the military R&D community (where possible and beneficial, in cooperation with the civilian community) in the following areas because, in the absence of large civilian markets, no one else is likely to support it (the inclusions in parentheses give examples of the kinds of capabilities and devices that would be included in each):

• Sensor technologies (electronically steered and low-probability-of-intercept (LPI) radar, IR and advanced infrared search and track (IRST), multispectral imaging, embedded microsensors and "smart" skins and structures, lasers, SQUIDs);

• The sensor technologies would be joined in application with specialized information technologies (secure data access; stealth and counterstealth; ASW; chemical, biological, and nuclear weapons detection; automatic target recognition) to contribute to the military parts of the information-in-warfare system. (Fundamental research into the theoretical basis of naval warfare underlying modeling and simulation must obviously be supported by the naval forces, as well.)

• Military-oriented materials (energetic materials, including explosives and rocket propellants, high-temperature materials for engine turbine blades and combustors, and composites, among others);

• The materials together with power and propulsion technologies (rocket engines, warheads, and advanced aircraft, ship, and submarine power plants) would contribute to the creation of advanced weapon systems and, in the form of long-life and high-power-density power sources, to reducing equipment loads and logistic resupply requirements.

In many of these areas, the naval forces will have to join with the other military departments to share the applied R&D and advanced development loads so that the total resources are spent as efficiently as possible. R&D expenditures by the Navy Department in these areas, and in the adaptation of civilian technology to naval force purposes, must be focused in two areas: development of unique naval force capabilities needed to support ongoing force improvement and creation of future capability; and development, by work-sharing arrangements in the joint environment, of capabilities that all the Services will be able to use. Deciding the allocation of resources between these two areas of effort will obviously be the responsibility of the Department of the Navy working with the Joint Chiefs of Staff, the other military departments, and the Office of the Secretary of Defense. Some of the jointly agreed R&D will help the naval forces, just as some of the Navy Department R&D will help meet needs of the other Services.

Within the Department of the Navy, the following areas of concentration for R&D application, associated with the entering wedges of capability and leading to their creation, should be especially fostered[25] (for completeness, the following list brings forward some critical R&D areas that were discussed in more detail in the report of the regional conflict study[26] than in this report; these items are starred in the list):

1. Information, intelligence, and space systems:
   — Information security, defensive information warfare;
   — *Satellite-based position-location security, deniability to opponents, within treaty commitments;
   — Penetration of concealment, cover, and deception for intelligence and situational awareness;
   — Preserving privacy, security, and military functionality while using commercial communications.

2. Human resources:
   — Distributed training;
   — Advanced casualty treatment and recovery, including chemical and biological casualty avoidance and treatment;
   — Data comprehension;
   — Quality-of-life research: QOL data collection; QOL metrics and analysis of return on investment in QOL.

3. Surface and air systems:
   — Rocket-propelled missile system design: staging and advanced, insensitive propellants for range extension, tailored warheads, terminal guidance, cold launch, at-sea reload, and cost reduction;
   — Target sensing, target recognition, and target location using unmanned platforms;
   — Continued work in stealth and counterstealth for all platforms, with special emphasis on the IR regime for aircraft signature reduction;
   — Continuation of ATBM systems development;
   — Laser weapons for ship defense against missiles, in the cooperative engagement capability (CEC) mode;
   — Electric systems, oriented toward advanced propulsion and power conditioning for Navy ships and submarines;

---

[25] There may be other areas of effort that are not mentioned in this list, that in the judgment of the Navy Department's R&D management have deserved program emphasis and resources. Failure to mention such an area of effort here does not carry the connotation that the study examined it and decided that it was of no importance, only that it was not directly connected with the entering wedges of capability described in this report.

[26] Naval Studies Board. 1996. *The Navy and Marine Corps in Regional Conflict in the 21st Century,* National Academy Press, Washington, D.C.

— Ship design for smaller crews—especially, distributed sensors, actuators, controls, and intelligent automated subsystems;

— Advanced ship hull forms, especially those contributing to speed, seakeeping, and stealth;

— Advanced submarine designs;

— Advanced combat aircraft design features, including more efficient, high thrust-weight ratio engines, lightweight unitary structures, microsensor-based aerodynamic flow control techniques, and low-speed aerodynamic and propulsion control techniques, to mitigate weight penalties associated with vertical or near-vertical lift;

— Advanced aircraft design and manufacturing processes, using simulation, electronic prototyping, and flexible tooling.

4. Undersea systems:

— Matched-field coherent processing technologies for extending passive ASW detection and tracking capability;

— Multistatic active ASW;

— Multispectrum active and passive nonacoustic sensors for both ASW and mine detection;

— Mobile underwater synoptic sensor networks;

— Ocean science and related technology developments;

— Secure tactical communications between undersea and surface, air, and space systems;

— Advanced explosives, undersea weapon warheads, and mine fusing and warheads;

— Ship defense against torpedoes;

— Advanced countermine warfare—rapid location and tagging, parallel neutralization, defeating "smart" minefields, explosive blasting of channels to the beach from the air with precision bomb emplacement and timing.

5. Ground forces and their combat support:

— *Target designation for precision weapon delivery on precisely known coordinates;

— *Reliable combat identification;

— *Integration with at-sea forces in the overall information and communication network, down to the smallest forward unit;

— *Reducing vulnerability of vertical-lift aircraft to shoulder-fired SAMs; airborne detection of minefields in landing zones;

— *Situation awareness, target detection, sensors, robotic vehicles, and nondestructive weaponry for fighting in built-up areas; techniques for operations other than war, nonlethal weapons, and crowd-control devices.

6. Logistics:
   — Design for readiness and minimal field maintenance;
   — *Adapting to fully containerized logistic support—packaging, transport, delivery, ships, airlift, depot handling, and "retail" distribution to forward units;
   — *Information-based logistic techniques, equipment, and systems for maintaining weapon system readiness and for delivering materiel to forces at sea and over the shore;
   — *Achieving sea-state 3 LOTS capability.

7. Modeling and simulation:
   — Military science and phenomenology;
   — Simulation science and methodology applicable to military systems;
   — Constructing and maintaining warfare-area and world databases;
   — Adopting and adapting related developments in civilian fields to military problems and activities;
   — Validating concepts and methodology.

Finally, it must be emphasized that some major system advances take place in major steps after ongoing research and advanced development have created new opportunities. This has been especially apparent in the aviation area, where ongoing R&D in propulsion, aerodynamics, and structures leads periodically to a major advance in capability embodied in a new class of aircraft. For this to happen, the R&D must be supported in a sustained, long-term program in which each step is built on the last, such that at significant points a new system can be built on the advances achieved to that time. An example is the Integrated High Performance Turbine Engine Technology (IHPTET) program, jointly sponsored by the Office of the Secretary of Defense (OSD), the Military Departments, and industry. This program, together with its predecessor Service programs, has led to major advances in turbine and compressor materials, advanced combustors and engine controls, and overall engine designs. These advances have led in turn to major improvements in thrust, thrust-weight ratio, and fuel economy, leading to the superior U.S. military aircraft engine performance we see today, and to significant advances in civilian aviation as well.

The areas of surface ship and submarine design and construction, ASW, and oceanography listed above need a similar model of integrated, sustained R&D support, with clearly defined goals and schedules, industry-government collaboration, and stable funding, to achieve the potential seen for them in this study.

8

Implications for the Department of the Navy

A CONCEPTUAL REVOLUTION

The future naval forces will have to be transformed into leaner forces (forces that have less redundancy and that depend critically on connections among diverse system elements) having more responsiveness, reach, and capability, while simultaneously sustaining the forces needed to meet ongoing national security needs. The Department of Defense, and the Department of the Navy within it, are exploring many avenues, including resource allocation among force size, readiness, and modernization; prioritization among new system acquisitions; and competitive privatization and outsourcing of services, to make the necessary resources available. It is beyond the scope of this study to evaluate these approaches or to explore new ones. It is appropriate, however, in recognition of the difficulty of the resource issue, to comment on the implications of resource management philosophy for the naval forces' evolution over the next 35 to 40 years, and for their ability to do what will be demanded of them in the security environment described earlier in this report.

The greater demands that will be made of naval forces in the coming decades, together with the relative scarcity of resources, will require a new conceptual basis for the design of the 21st-century naval forces. New technology will open opportunities to provide those forces the capabilities described earlier, but only if the technology is applied according to the new formulation of principles for investment.

It has already been accepted in naval force planning that the forces will have to substitute capital for labor, using instrumentation, automation, and capability-multiplying technology, from computers to complex systems that need fewer people for control. In addition, most future force plans accept the need to substitute quick response, reach, and precision for numbers, by using information, speed, range, responsiveness, and weapon guidance to require fewer engagements per target and thereby allow smaller forces to accomplish military missions that have been assigned to large forces in the past. Carried further, this implies substituting efficiency, precision, and effectiveness for brute force in military operations. Information warfare and what the Joint Chiefs of Staff (JCS) "Vision 2010" calls "dominant maneuver" and "focused logistics" will have to be used to bring U.S. naval forces to points of decision to impose their will in crisis or conflict before they can be thwarted by any opposition. Assuming timely decision making by the appropriate government authorities, being at the right place at the right time with the right tools to eliminate the opponent's ability to fight will be far better than taking on an opponent with massive accumulations of force in areas and circumstances where the opponent has had time to build great strength.

Finally, planning resource use to create the forces will require joining value with dollars in thinking about expenditures. The naval forces are already thinking in the direction of designing for smaller crews, systems needing less support, and utilization of commercial services for many functions, to get more value for the dollar. In the future, life-cycle costs rather than acquisition costs will have to govern decisions about expenditures, in recognition that reduction of system support costs will make more resources available for continual force modernization and recapitalization within given budgets. System acquisition costs will have to be viewed as investments in capability with payoff over the long term rather than as purchases of individual platforms or weapons. In this approach, "affordability" must come to mean purchasing needed value for the money the Navy Department is willing and able to spend for a capability within its allocated budget, rather than simply spending the least amount of money in any area, as the term has come to be used in many parts of the Defense Department.

PAYOFFS AND VULNERABILITIES

The restructured naval forces that would emerge after such changes in thinking about naval force design, and after integration of the new capabilities described in the previous chapter, would be leaner and more powerful than today's forces, and able to do more within a given budget. They would be capable of responding more rapidly to crises, a capability enabled by power projection from farther out at sea to deeper inland by a greater variety of forces. Moreover, they would be capable of accommodating their response to a wider variety of crises that may range from invasion of an ally's territory to containing and reversing the effects of civil disturbance or terrorist action that threatens U.S. interests. The restructured forces would enable a more precise focus on the critical aspects of crises requiring combat or other operations, leading to earlier success in action. This would lead, in turn, to fewer naval force and allied casualties, less damage to the forces' major platforms and fewer losses of major equipment, fewer coincidental casualties among local populations, and less collateral damage. The transformation of the forces would bring with it a revised, more flexible cost structure for the naval forces, making continual modernization easier to sustain in the face of the rapidly evolving and spreading world technology base. It is apparent that benefits in these directions would increase as the rate of evolution increases from today's naval forces to those visualized for future decades.

The benefits cannot be achieved, however, without incurring some serious vulnerabilities, which will have to be dealt with. The following list describes those that will be the most difficult to deal with, along with some indications of means to mitigate their potential effects:

- *The forces would be heavily dependent on their communications and information structure, much of it commercial.* To mitigate the risk of interruption in information flow, the forces will have to practice smart usage: defensive information warfare, many redundant links via commercial as well as military systems, and antijamming and protective electronic warfare where essential.

- *Virtually all electronics (in sensors, communications, weapons, platforms) will continue to be vulnerable to destruction by the electromagnetic pulse (EMP) that would attend a nuclear burst, or that could be generated as part of a deliberate electronic warfare campaign.* The only certain protection is in hardening the electronic circuits, a measure that has been foregone in the past in other than special circumstances because it entails substantial added costs, and that may not be justifiable for military systems or available for commercial systems for that reason. Some advanced microcircuit materials will be inherently resistant to EMP. The extent to which such materials will be used in systems that depend heavily on commercial equipment and devices is problematic. There may be some protection in the fact that commercial communications will likely be shared with opponents who have similar access in the global economy.

- *Information systems will be subject to defeat by concealment, cover, and deception.* There will be protection in multisource data input and correlation, multispectral imaging, foliage-penetrating radar, and greater use of human intelligence inputs. It will be vital for our own naval force commanders to learn and understand potential opponents' culture and habits of thinking as part of their own "kit of tools," to gain insights into the potential directions for surprise and deception that a particular opponent may pose.

- *Unmanned systems operating autonomously to deliver weapons could misidentify targets, causing undesirable consequences or even tragedy.* There must be a doctrine and "rules of engagement" governing the operation of autonomous unmanned platforms for weapon delivery, means for monitoring their performance on missions, and intervention to prevent unwanted damage and outcomes.

- *The "lean" organization of the future naval forces, characterized by a lack of redundancy and the need for actions relying on smooth and accurate transfer of up-to-date information, could be brittle under the fog and friction of war.* Decentralized command and control, with more authority and responsibility at lower echelons of the force and more complete situational awareness and secure combat identification, can help guard against this potential fragility. Sturdy communications are the critical element in creating such safeguards. Another key problem posed by lean organizations and operations is the need for fallback positions in case plans go awry and essential force elements are not in place as expected. Nowhere will this be more critical than in the information area supporting the maintenance of situational awareness and targeting for the naval forces if some of the joint assets, such as the Joint Surveillance and Target Attack Radar System (JSTARS), are not available when needed. Review of the potential condition of the naval forces in such a case shows that with the information from space to which the naval forces will have full access, advances in the capability of the E-2C system, reconnaissance pods that can be flown by fighter aircraft, UAVs that the naval forces will have and will operate, and planned communications links, there should be sufficient capability to operate effectively until the full joint system can be brought into place or reconstituted if it is interrupted. The naval forces must, however, take steps to ensure that the forces have the minimal capability needed for fully independent operations. This may require, for example, creating a simplified JSTARS-like capability to locate and identify opposing ground forces and targets, and the ability to locate any targets found with naval force assets in the common GPS grid and universal time.

- *"Lean" forces would be postured for quick victory; an opponent might outlast them.* The chief protection against such an outcome lies in the fact that more of the opposition would be engaged by more of our forces at points critical to the opposition's defeat; this should hasten that defeat. Also, in case an opponent's staying power requires rapid expansion of our forces, our forces would be better postured for expansion in appropriate directions than they are now if they embodied the new capabilities described in this report. Residual opposition forces in a conflict to secure an ally's territory might undertake guerrilla or terrorist warfare. It would be the U.S. task to help the ally deal with such an outcome expeditiously; the restructured naval forces would be better able to do that than are today's forces.

- *The naval forces visualized for the future would be attuned to a high-technology opponent, with capabilities based on a level of technology roughly equivalent to our own; a low-technology or no-technology opponent—for example, one who communicates by non-electronic means—could pose problems the future systems would not be designed to handle.* To meet this contingency, our forces will have to understand potential opponents by preparing in advance for likely areas of engagement, and appropriately training the lower-level troop commanders who will likely be the first to encounter unexpected tactics and techniques. Our expeditionary forces will also be able to arrange for distributed area expertise on call, and to take advantage of coalition partners' knowledge of the opposition.

- *Future naval forces will plan and train extensively with M&S in a "virtual world"; they could eventually lose touch with the real world.* The only way to guard against this and still have the M&S support is to undertake continual field exercises intermixed with the simulations, preferably in a joint and combined environment, and to continually test M&S results against real-life situations and history to ensure that they do not unintentionally depart too far from reality.
- *Extensive use of commercial support could lead to reduced military control of key support elements of the forces in time of crisis, consequently interfering with the forces' ability to perform their tasks—for example, by work stoppages or dilution and diversion of resources.* This would be a national, not solely a naval force, problem. The problem was faced in World War II and resolved by special acts of Congress. For the future, too, the DOD and Congress must establish and enforce "rules of the game" appropriate to the new designs of the armed forces and their support structure.

No military force or national effort using that force can be entirely free of vulnerability to opposition actions. As the notes above suggest, prudent steps can be taken to mitigate the worst effects of the vulnerabilities that would face newly designed naval forces. Such mitigation efforts must be undertaken as the new capabilities are built, as part of the system and force design, and their cost must be considered an integral part of the cost of implementing new naval force capabilities.

IMPLICATIONS FOR NAVAL FORCE PLANNING

The naval forces are currently shrinking. The decision to expand them will most likely be made when there appears on the horizon a substantially more serious threat to our national security than we perceive today. Judging from past history, such circumstances will not allow the luxury of extensive and time-consuming experimentation with new kinds of systems, forces, and concepts of operation. The expanded naval forces will be built on the foundation of the forces and capability that exist at the time.

If modernization before that need appears remains cautious and fractionated in the budget squeeze, there will be low technical risk but a high risk of technical and operational obsolescence vis-à-vis any emerging threat. The naval forces will retain largely the same characteristics as today's forces, and the budget structure could lock in current manpower-intensive systems for a long time. Roughly today's kind of naval forces, with limited improvements and efficiencies, would continue when expansion is needed. These forces may not be well positioned to meet new kinds of threat that are likely to emerge.

If the entering wedges of transformation are pursued aggressively, technical, financial, and operational risk will be higher in many areas, even though the risk could be minimized by incremental and evolutionary approaches to introduction and evaluation of major innovations. However, the new kinds of naval forces that emerge would be far better positioned to adapt and meet new kinds of threats to national security when expansion is needed. To achieve this position, early commitment to many new and challenging concepts would be required, with the risk of cost growth, delay, or failure in some of the new directions that would not be tolerated easily under stringent budget conditions. For this reason, and because such budget conditions increase the likelihood that resources would be lost to the naval forces when resources are shifted from one area of effort to another, a broad base of support is needed for the transformation throughout the Defense Department, the Executive Branch, and the Congress. Difficult and uncertain though the process and the outcome must be, the naval forces' forward posture and potential for earliest engagement require the Department of the Navy to build that support as part of the process of naval force restructuring.

AN EVOLUTIONARY APPROACH TO REVOLUTIONARY CAPABILITY

Many explorations of new technical and operational directions are under way in the naval forces—in approaches to using information in warfare, in the emerging Marine Corps Operational Maneuver From the Sea doctrine and concepts of operation, in personnel management, in ships, aircraft, submarines, weapons, and their employment and logistic support, and in joint operations and usage. These new directions, which imply radical change in the future naval forces, have already begun to create the entering wedges of capability upon which future naval forces will be built. The emerging capabilities must be tested operationally in the forces and their ultimate development guided in directions that will ensure their viability. When these directions are determined, the new capabilities must then be joined with existing long-term investments in C^4ISR systems, weapon systems, and platforms that will remain useful in any kind of naval force for years and decades to come, in an evolutionary approach to restructured naval forces.

One such evolutionary approach is illustrated in Figure 8.1. The figure shows the decades between 2000 and 2040 during which many existing weapon systems and platforms will reach the end of their service life (ESL), and during which replacements embodying the new capabilities could enter the forces. The implementation schedule shown is not a "hard and fast" recommendation, but illustrative. It recognizes that some investments, such as those in major ships like aircraft carriers and a generation of combat aircraft, have very long service lives, and that weapon systems, like the family of attack ballistic missiles de-

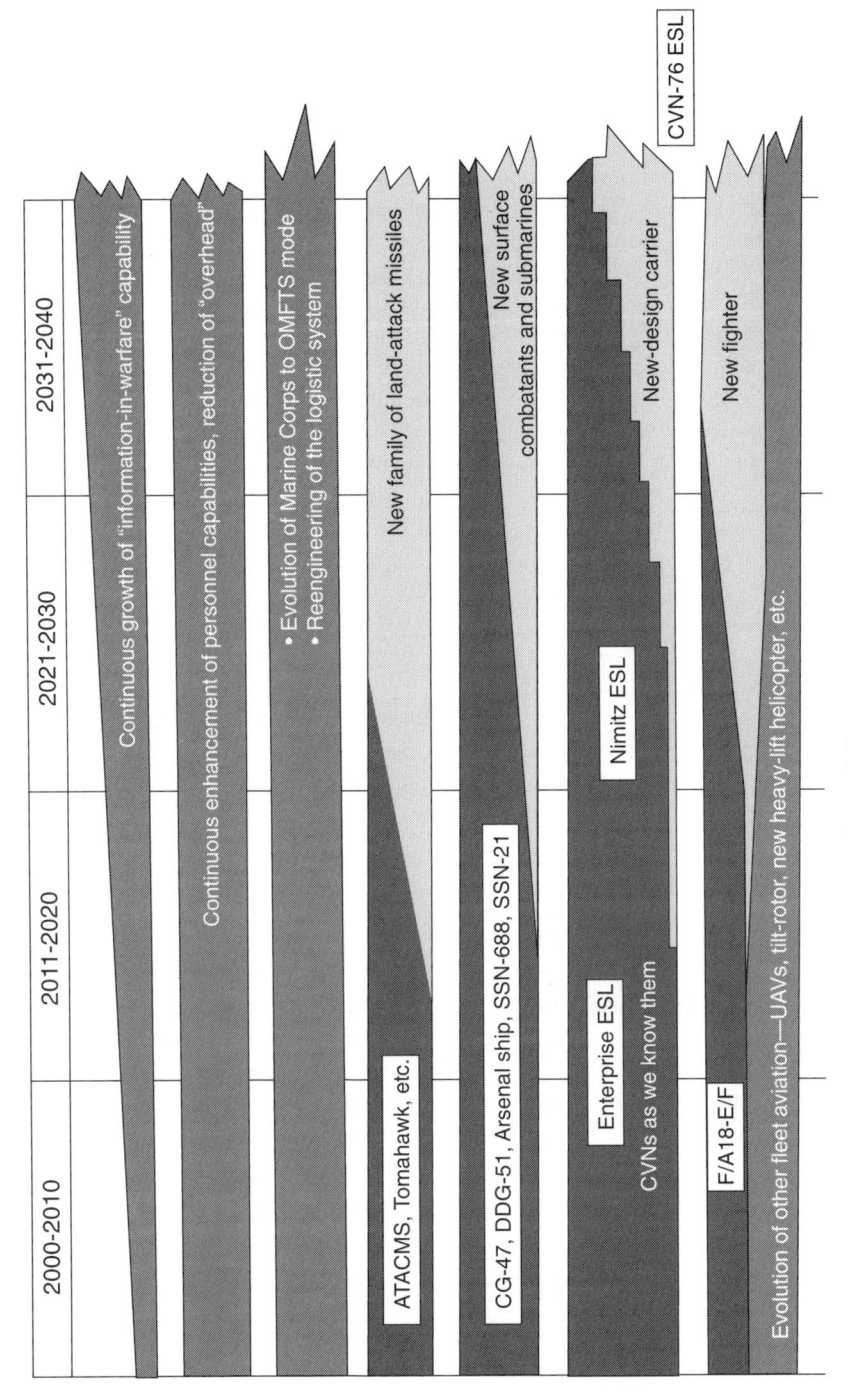

FIGURE 8.1 Evolutionary path to restructured naval forces.

scribed previously, will take time to develop with all the technical characteristics that advance them significantly beyond today's weapon systems.

Information capabilities are developing much more rapidly than platforms and weapons can be developed. Personnel and financial management that capitalizes on available technology can be changed significantly in relatively short periods of time. Successive advances in these areas can be integrated into the forces at any stage of evolution of the major hardware systems that take longer to create (where the term "hardware" refers to any durable parts of naval force systems). Conversely, the new hardware systems will be able to take advantage of the advances in the information, personnel, and management areas, and they will be designed to do so. Improvements in the logistic system would occupy an intermediate position, since although conceptual changes can be made rapidly, it will take time to implement some of the hardware and the software process changes needed. Similarly, changes in doctrine and concepts of operation will show the way to the hardware developments needed, but will also have to await the hardware availability for full implementation.

The illustration demonstrates that over the time period covered by this study revolutionary change in the structure and capability of the naval forces can be achieved by a manageable evolutionary path. The resulting forces will be more capable and more adaptable to the unexpected challenges posed by an uncertain world than are today's forces, warranting the risks entailed in starting down the pathway to such extensive change.

It is difficult to see very far into the future of developing technologies. The study group has been aware that an effort like this one, undertaken at the beginning of the 20th century, would not have predicted two world wars, with one of them using many thousands of aircraft and massed amphibious landings as controlling elements, "in the next 40 years."

Nor would a study at the end of that war have foreseen in 1945 a strategic weapons balance between world powers, based on nuclear-powered submarines loaded with intercontinental-range nuclear missiles, in another third of a century.

The last third of the current century has given us computer technology and space systems—both even yet of uncertain but surely large impact in the future. If that future has as many "impossible" advances waiting to appear, it surely seems wise for the Navy and Marine Corps to continue an examination of the technological future every decade or so. The members of this study have enjoyed the current exercise in this spirit.

APPENDIXES

A

Terms of Reference

CHIEF OF NAVAL OPERATIONS

28 November 1995

Dear Dr. Alberts,

   In 1986, at the request of this office, the Academy's Naval Studies Board undertook a study entitled "Implications of Advancing Technology for Naval Warfare in the Twenty-First Century." The Navy-21 report, as it came to be called, projected the impact of evolving technologies on naval warfare out to the year 2035, and has been of significant value to naval planning over the intervening years. However, as was generally agreed at the time, the Navy and Marine Corps would derive maximum benefit from a periodic comprehensive review of the implications of advancing technology on future Navy and Marine Corps capabilities. In other words, at intervals of about ten years, the findings should be adjusted for unanticipated changes in technology, naval strategy, or national security requirements. In view of the momentous changes that have since taken place, particularly with national security requirements in the aftermath of the Cold War, I request that the Naval Studies Board immediately undertake a major review and revision of the earlier Navy-21 findings.

   The attached Terms of Reference, developed in consultation between my staff and the Chairman and Director of the Naval Studies Board, indicate those topics which I believe should receive special attention. If you agree to accept this request, I would appreciate the results of the effort in 18 months.

Sincerely,

J. M. BOORDA
Admiral, U.S. Navy

Dr. Bruce M. Alberts
President
National Academy of Sciences
2101 Constitution Avenue, N.W.
Washington, D.C. 20418

Enclosure

APPENDIX A

TERMS OF REFERENCE

TECHNOLOGY FOR THE FUTURE NAVY

The Navy-21 study (Implications of Advancing Technology for Naval Warfare in the Twenty-First Century), initiated in 1986 and published in 1988, projected the impact of technology on the form and capability of the Navy to the year 2035. In view of the fundamental national and international changes -- especially the Cold War's end -- that have occurred since 1988, it is timely to conduct a comprehensive review of the Navy-21 findings, and recast them, where needed, to reflect known and anticipated changes in the threat, naval missions, force levels, budget, manpower, as well as present or anticipated technical developments capable of providing cost effective leverage in an austere environment. Drawing upon its subsequent studies where appropriate, including the subpanel review in 1992 of the prior Navy-21 study, the Naval Studies Board is requested to undertake immediately a comprehensive review and update of its 1988 findings. In addition to identifying present and emerging technologies that relate to the full breadth of Navy and Marine Corps mission capabilities, specific attention also will be directed to reviewing and projecting developments and needs related to the following: (1) information warfare, electronic warfare, and the use of surveillance assets; (2) mine warfare and submarine warfare; (3) Navy and Marine Corps weaponry in the context of effectiveness on target; (4) issues in caring for and maximizing effectiveness of Navy and Marine Corps human resources. Specific attention should be directed, but not confined to, the following issues:

1. Recognizing the need to obtain maximum leverage from Navy and Marine Corps capital assets within existing and planned budgets, the review should place emphasis on surveying present and emerging technical opportunities to advance Navy and Marine Corps capabilities within these constraints. The review should include key military and civilian technologies that can affect Navy and Marine Corps future operations. This technical assessment should evaluate which science and technology research must be maintained in naval research laboratories as core requirements versus what research commercial industry can be relied upon to develop.

2. Information warfare, electronic warfare and the exploitation of surveillance assets, both through military and commercial developments, should receive special attention in the

review. The efforts should concentrate on information warfare, especially defensive measures that affordably provide the best capability.

3. Mine warfare and submarine warfare are two serious threats to future naval missions that can be anticipated with confidence, and should be treated accordingly in the review. This should include both new considerations, such as increased emphasis on shallow water operations, and current and future problems resident in projected worldwide undersea capability.

4. Technologies that may advance cruise and tactical ballistic missile defense and offensive capabilities beyond current system approaches should be examined. Counters to conventional, bacteriological, chemical and nuclear warheads should receive special attention.

5. The full range of Navy and Marine Corps weaponry should be reviewed in the light of new technologies to generate new and improved capabilities (for example, improved targeting and target recognition).

6. Navy and Marine Corps platforms, including propulsion systems, should be evaluated for suitability to future missions and operating environments. For example, compliance with environmental issues is becoming increasingly expensive for the naval service and affects operations. The review should take known issues into account, and anticipate those likely to affect the Navy and Marine Corps in the future.

7. In the future, Navy and Marine Corps personnel may be called upon to serve in non-traditional environments, and face new types of threats. Application of new technologies to the Navy's medical and health care delivery systems should be assessed with these factors, as well as joint and coalition operations, reduced force and manpower levels, and the adequacy of specialized training in mind.

8. Efficient and effective use of personnel will be of critical importance. The impact of new technologies on personnel issues, such as education and training, recruitment, retention and motivation, and the efficient marriage of personnel and machines should be addressed in the review. A review of past practices in education and training would provide a useful adjunct.

APPENDIX A

9. Housing, barracks, MWR facilities, commissaries, child care, etc. are all part of the Quality of Life (QOL) of naval personnel. The study should evaluate how technology can be used to enhance QOL and should define militarily meaningful measures of effectiveness (for example, the impact on Navy readiness).

10. The naval service is increasingly dependent upon modeling and simulation. The study should review the overall architecture of models and simulation in the DoD (DoN, JCS, and OSD), the ability of models to represent real world situations, and their merits as tools upon which to make technical and force composition decisions.

The study should take 18 months and produce a single-volume overview report supported by task group reports (published either separately or as a single volume). Task group reports should be published as soon as completed to facilitate incorporation into the DoN planning and programming process. An overview briefing also should be produced that summarizes the contents of the overview report, including the major findings, conclusions, and recommendations.

B

Study Membership and Participants

COMMITTEE ON TECHNOLOGY FOR FUTURE NAVAL FORCES

DAVID R. HEEBNER, Science Applications International Corporation (retired) *Study Chair*
ALBERT J. BACIOCCO, JR., The Baciocco Group, *Chair, Panel on Undersea Warfare*
ALAN BERMAN, Applied Research Laboratory, Pennsylvania State University, *Chair, Panel on Weapons*
NORMAN E. BETAQUE, Logistics Management Institute, *Chair, Panel on Logistics*
GERALD A. CANN, Raytheon Company, *Chair, Panel on Platforms*
GEORGE F. CARRIER, Harvard University, *Chair, Panel on Modeling and Simulation*
SEYMOUR J. DEITCHMAN, Institute for Defense Analyses (retired), *Chair, Coordination and Integration*
ALEXANDER FLAX, Potomac, Maryland, *Senior Advisor*
WILLIAM J. MORAN, Redwood City, California, *Senior Advisor*
ROBERT J. MURRAY, Center for Naval Analyses, *Chair, Panel on Human Resources*
ROBERT B. OAKLEY, National Defense University, *Senior Advisor*
JOSEPH B. REAGAN, Saratoga, California, *Chair, Panel on Technology*
VINCENT VITTO, Lincoln Laboratory, Massachusetts Institute of Technology, *Chair, Panel on Information in Warfare*

APPENDIX B

CPSMA Liaisons
JOHN E. ESTES, University of California at Santa Barbara
KENNETH H. KELLER, University of Minnesota
W. CARL LINEBERGER, University of Colorado
CHARLES A. ZRAKET, Mitre Corporation (retired)

Navy Liaisons
RADM JOHN W. CRAINE, JR., USN, Office of the Chief of Naval Operations, N81 (as of July 4, 1996)
VADM THOMAS B. FARGO, USN, Office of the Chief of Naval Operations, N81 (through July 3, 1996)
RADM RICHARD A. RIDDELL, USN, Office of the Chief of Naval Operations, N91
CDR DOUGLASS BIESEL, USN, Office of the Chief of Naval Operations, N812C1

Marine Corps Liaison
LtGen PAUL K. VAN RIPER, USMC, Marine Corps Combat Development Command

ADVISORY COUNCIL

ROBERT A. FROSCH, John F. Kennedy School of Government, Harvard University, *Chair*
MALCOLM R. CURRIE, Hughes Aircraft Company (Chairman Emeritus)
RUTH M. DAVIS, Pymatuning Group, Inc.
JOHN S. FOSTER, TRW, Inc.
ALFRED M. GRAY, Alexandria, Virginia
WILLIS M. HAWKINS, Woodland Hills, California
ROBERT L.J. LONG, Annapolis, Maryland
ROBERT W. LUCKY, Bell Communications Research, Inc.
LARRY D. WELCH, Institute for Defense Analyses

CPSMA Liaison
ROBERT J. HERMANN, United Technologies Corporation

PANEL ON TECHNOLOGY

JOSEPH B. REAGAN, Saratoga, California, *Chair*
HERBERT RABIN, University of Maryland, *Vice Chair*
SUSAN D. ALLEN, Florida State University
RONALD CLARK, Lockheed Martin Corporation
ANTHONY J. DeMARIA, DeMaria ElectroOptics Systems, Inc.

DANIEL N. HELD, Northrop Grumman Corporation
RAY L. LEADABRAND, Leadabrand and Associates, Inc.
DAVID W. McCALL, Far Hills, New Jersey
WILLIAM J. PHILLIPS, Northstar Associates, Inc.
DENNIS L. POLLA, University of Minnesota
MARA G. PRENTISS, Harvard University
JOHN W. ROUSE, JR., Southern Research Institute
ALBERT I. SCHINDLER, Rockville, Maryland
STEVEN J. TEMPLE, Raytheon Company
EDWARD W. THOMPSON, Hughes Research Laboratory
ROBERT M. WESTERVELT, Harvard University

Invited Participants
ARISTOS CHRISTOU, University of Maryland
FRANK A. HORRIGAN, Raytheon Electronic Systems
JOHN W.R. POPE, JR., Tri-Tech Microproducts
TIMOTHY D. ROARK, TRW
HOWARD STEVENS, Vector Research (as of January 1, 1997)

Navy Liaison Representatives
PAUL G. BLATCH, Office of the Chief of Naval Operations, N911T1
FRED WOLPERT, Office of the Chief of Naval Operations, N911T2

PANEL ON INFORMATION IN WARFARE

VINCENT VITTO, Lincoln Laboratory, Massachusetts Institute of Technology, *Chair*
PHILIP S. ANSELMO, Northrop Grumman Corporation, *Vice Chair*
NORVAL L. BROOME, Mitre Corporation
J. ROBERT COLLINS, E Systems
BURTON I. EDELSON, George Washington University
JOHN F. EGAN, Lockheed Martin Corporation
ROBERT HUMMEL, Courant Institute of Mathematical Sciences, New York University
GERALD McNIFF, Northrop Grumman Corporation
ROBERT NESBIT, Mitre Corporation
STANLEY R. ROBINSON, Environmental Research Institute of Michigan
JULIE JCH RYAN, Booz, Allen and Hamilton
H. GREGORY TORNATORE, Applied Physics Laboratory, Johns Hopkins University
BRUCE WALD, Center for Naval Analyses
MARY LETICIA VAJTA-WILLIAMS, Space Imaging, Inc.

APPENDIX B

Navy Liaison Representatives
LCDR HARRY COKER, USN, Department of Defense Space Architect
CAPT MATTHEW ROGERS, USN, Department of Defense Space Architect
LtCol FRANK WALIZER, USMC, Office of the Chief of Naval Operations, N853H
CAPT MICHAEL WINSLOW, USN, Office of the Chief of Naval Operations, N6C

PANEL ON HUMAN RESOURCES

ROBERT J. MURRAY, Center for Naval Analyses, *Chair*
J. DEXTER FLETCHER, Institute for Defense Analyses, *Vice Chair*
PAUL R. CHATELIER, Defense Modeling and Simulation Office
DONALD J. CYMROT, Center for Naval Analyses
WARREN S. GRUNDFEST, Cedars-Sinai Medical Center
LEE D. HEIB, Yuma, Arizona
ELYSE W. KERCE, Madison, Alabama
REUVEN LEOPOLD, Syntek
R. BOWEN LOFTIN, University of Houston
JAMES A. MARTIN, Bryn Mawr College
JOSEPH M. ROSEN, Dartmouth-Hitchcock Medical Center
HARRISON SHULL, Monterey, California
NORMAN H. SMITH, Linden, Virginia
J. PACE VanDEVENDER, Prosperity Institute
JOSEPH ZEIDNER, George Washington University

Navy Liaison Representatives
CDR MARK BURGUNDER, USN, Office of the Chief of Naval Operations, N752
SANDRA CRUM, Office of the Chief of Naval Operations, N464
CDR CHARLES ENGLISH, USN, Office of the Chief of Naval Operations, N931
CDR GILBERT GIBSON, USN, Office of the Chief of Naval Operations, N971P
CDR CHARLES GUNN, USN, Bureau of Medicine
LCDR WILLIAM JONSON, USN, Office of the Chief of Naval Operations, N752
CDR BARBARA KOROSEC, USN, Office of the Chief of Naval Operations, N122B
CDR DAVID SORANNO, USN, Office of the Chief of Naval Operations, N712
DR. DANIEL STABILE, Office of the Chief of Naval Operations, PERS6D

PANEL ON WEAPONS

ALAN BERMAN, Applied Research Laboratory, Pennsylvania State University, *Chair*
GEORGE S. SEBESTYEN, McLean, Virginia, *Vice Chair*
VICTOR C.D. DAWSON, Poolesville, Maryland
NORMAN E. EHLERT, Gig Harbor, Washington
MAURICE EISENSTEIN, RAND Corporation
MILTON FINGER, Lawrence Livermore National Laboratory
RAY "M" FRANKLIN, Port Angeles, Washington
JACK E. GOELLER, Advanced Technology Research Corporation
ALFRED B. GSCHWENDTNER, Lincoln Laboratory, Massachusetts Institute of Technology
FRANK KENDALL, Lexington, Massachusetts
IRA F. KUHN, Directed Technologies, Inc.
DIANA F. McCAMMON, Applied Research Laboratory, Pennsylvania State University
CHARLES F. SHARN, McLean, Virginia
WALTER SOOY, Pleasanton, California
VERENA S. VOMASTIC, Institute for Defense Analyses

Navy Liaison Representatives
CDR THOMAS COSGROVE, USN, Office of the Chief of Naval Operations, N858D
CAPT JOHN McGILLVRAY, USN, Office of the Chief of Naval Operations, N863J
CDR DENNIS MURPHY, USN, Office of the Chief of Naval Operations, N87C1
LCDR PETE McSHEA, USN, Office of the Chief of Naval Operations, N88W3

PANEL ON PLATFORMS

GERALD A. CANN, Raytheon Company, *Chair*
WILLIAM D. O'NEIL, Center for Naval Analyses, *Vice Chair*
STEVEN D. ADAMS, Bath Iron Works Corporation
JAMES P. BROOKS, Litton/Ingalls Shipbuilding, Inc.
ROY R. BUEHLER, Lockheed Martin Aeronautical Systems
DANIEL L. COOPER, Wyomissing, Pennsylvania
RICHARD M. DUNLEAVY, Virginia Beach, Virginia
STANLEY F. DUNN, Florida Atlantic University
ROBERT H. GORMLEY, The Oceanus Company
JAMES C. HAY, Potomac, Maryland

APPENDIX B

CHARLES F. HORNE III, Mount Pleasant, South Carolina
ANN R. KARAGOZIAN, University of California at Los Angeles
RONALD K. KISS, Rockville, Maryland
THOMAS C. MALONEY, General Dynamics
DAVID W. McCALL, Far Hills, New Jersey
IRWIN MENDELSON, Singer Island, Florida
JOSEPH METCALF III, Washington, D.C.
RICHARDS T. MILLER, Annapolis, Maryland
MICHAEL L. POWELL, Newport News Shipbuilding, Inc.
JAMES M. SINNETT, McDonnell Douglas Corporation
KEITH A. SMITH, Vienna, Virginia
JAMES J. TURNER, SR., Alexandria, Virginia

Invited Participant
MICHAEL T. BOYCE, Boeing Defense and Space Group

Navy Liaison Representatives
CAPT JOHN McGILLVRAY, USN, Office of the Chief of Naval Operations, N863J
CDR DENNIS MURPHY, USN, Office of the Chief of Naval Operations, N87C1
LCDR PETE McSHEA, USN, Office of the Chief of Naval Operations, N88W3

PANEL ON UNDERSEA WARFARE

ALBERT J. BACIOCCO, JR., The Baciocco Group, Inc., *Chair*
RICHARD F. PITTENGER, Woods Hole Oceanographic Institution, *Vice Chair*
JOHN W. ASHER III, Global Associates, Ltd.
ARTHUR B. BAGGEROER, Massachusetts Institute of Technology
DAVID B. BURKE, JR., Charles Stark Draper Laboratory, Inc.
NICHOLAS P. CHOTIROS, Applied Research Laboratory, University of Texas at Austin
MYRON P. GRAY, Applied Physics Laboratory, Johns Hopkins University
THOMAS C. HALSEY, Exxon Research and Engineering Company
RICHARD F. HOGLUND, King George, Virginia
ERNEST L. HOLMBOE, Applied Physics Laboratory, Johns Hopkins University
WILLIAM J. HURLEY, Institute for Defense Analyses
WESLEY E. JORDAN, Bolt, Beranek and Newman Co.
CECIL J. KEMPF, Coronado, California

EDWARD G. LISZKA, Applied Research Laboratory, Pennsylvania State University
R. KENNETH LOBB, Applied Research Laboratory, Pennsylvania State University
DANIEL M. NOSENCHUCK, Princeton University
THOMAS D. RYAN, Nuclear Energy Institute
KEITH A. SMITH, Vienna, Virginia
ROBERT C. SPINDEL, Applied Physics Laboratory, University of Washington
DAVID L. STANFORD, Science Applications International Corporation

Invited Participants
IRA DYER, Massachusetts Institute of Technology
WILLIAM A. KUPERMAN, Scripps Institution of Oceanography

Navy Liaison Representatives
CDR THOMAS COSGROVE, USN, Office of the Chief of Naval Operations, N858D
CAPT JOHN McGILLVRAY, USN, Office of the Chief of Naval Operations, N863J
CDR DENNIS MURPHY, USN, Office of the Chief of Naval Operations, N87C1
LCDR PETE McSHEA, USN, Office of the Chief of Naval Operations, N88W3
ALLISON STILLER, Office of the Assistant Secretary of the Navy, RDA

PANEL ON LOGISTICS

NORMAN E. BETAQUE, Logistics Management Institute, *Chair*
PHILIP D. SHUTLER, Annandale, Virginia, *Vice Chair*
ANDREW A. GIORDANO, The Giordano Group, Ltd.
MICHAEL P. KALLERES, Global Associates, Ltd.
DAVID B. KASSING, RAND Corporation
PAUL A. LAUTERMILCH, Clarkesville, Virginia
IRWIN MENDELSON, Singer Island, Florida
DANIEL SAVITSKY, Davidson Laboratory, Stevens Institute of Technology
W. HUGH WOODIN, University of California at Berkeley

Navy Liaison Representatives
CDR THOMAS COSGROVE, USN, Office of the Chief of Naval Operations, N858D
CDR ERIC FERRARO, USN, Office of the Chief of Naval Operations, N62M
NICHOLAS M. LINKOWITZ, Marine Corps Combat Development Command

APPENDIX B

LCDR PETER MORGAN, USN, Office of the Chief of Naval Operations, N422F
LCDR PAT THURMAN, USN, Office of the Chief of Naval Operations, N432II

PANEL ON MODELING AND SIMULATION

GEORGE F. CARRIER, Harvard University, *Chair*
PAUL K. DAVIS, RAND Corporation, *Vice Chair*
DONALD K. BLUMENTHAL, Gualala, California
RICHARD BRONOWITZ, Center for Naval Analyses
JOHN C. DOYLE, California Institute of Technology
DONALD P. GAVER, Naval Postgraduate School
DON E. HIHN, Charleston, South Carolina
RICHARD J. IVANETICH, Institute for Defense Analyses
JOHN P. LEHOCZKY, Carnegie Mellon University
DAVID L. McDOWELL, Georgia Institute of Technology
DUNCAN C. MILLER, Massachusetts Institute of Technology
DAVID R. OLIVER, Northrop Grumman Corporation
GABRIEL ROBINS, University of Virginia
BERNARD P. ZIEGLER, University of Arizona

Invited Participant
BEN P. WISE, Science Applications International Corporation

Navy Liaison Representatives
CDR THOMAS COSGROVE, USN, Office of the Chief of Naval Operations, N858D
CAPT JAY KISTLER, USN, Office of the Chief of Naval Operations, N6M

CONSULTANTS

LEE M. HUNT
SIDNEY G. REED, JR.
JAMES G. WILSON

NRC STAFF

RONALD D. TAYLOR, Director
PETER W. ROONEY, Program Officer
SUSAN G. CAMPBELL, Administrative Assistant
MARY G. GORDON, Information Officer
CHRISTOPHER A. HANNA, Project Assistant

C

Acronyms and Abbreviations

AAW	Antiair warfare
AEW	Airborne early warning
AIP	Air-independent propulsion
AMCM	Airborne mine countermeasures
ARG	Amphibious ready group
ASUW	Antisurface warfare
ASW	Antisubmarine warfare
ATACMS	Army Tactical Missile System
ATBM	Antitactical or antitheater ballistic missile
BFTT	Battle Force Tactical Trainer
BUR	Bottom Up Review
BW/CW	Biological warfare/chemical warfare
C^3I	Command, control, communications, and intelligence
C^4ISR	Command, control, communications, computing, intelligence, surveillance, and reconnaissance
CEC	Cooperative engagement capability
CEP	Circular error of probability
CIA	Central Intelligence Agency
CINC	Commander-in-chief
CLZ	Craft landing zone
CONUS	Continental United States
COTS	Commercial off the shelf
CRAF	Civil Reserve Airlift Fleet
DARO	Defense Aerial Reconnaissance Office

APPENDIX C

DARPA	Defense Advanced Research Projects Agency
DET	Explosive net (mine warfare)
DIA	Defense Intelligence Agency
DIS	Distributed interactive simulation
DOD	Department of Defense
EER	Explosive echo ranging
ELINT	Electronic intelligence
EMCON	Emission control
EMNV	Expendable mine neutralization vehicle
EMP	Electromagnetic pulse
EOD	Explosive ordnance disposal
ERGM	Extended-range guided munition
ESL	End of service life
ESM	Electronic support measures
ESSM	Evolved Sea Sparrow missile
EW	Electronic warfare
FARP	Forward arming and refueling point
GPS	Global Positioning System
ICBM	Intercontinental ballistic missile
IHPTET	Integrated High Performance Turbine Engine Technology (program)
IR	Infrared
IRST	Infrared search and track
ISR	Intelligence, surveillance, reconnaissance
JCS	Joint Chiefs of Staff
JSIMS	Joint Staff Simulation System
JSTARS	Joint Surveillance and Target Attack Radar System
JWARS	Joint Warfare System
LFA	Low-frequency active
LIDAR	Light detection and ranging
LOTS	Logistics (delivery) over the shore
LPI	Low probability of intercept
MARSIM	Maritime operation simulation
MCM	Mine countermeasures
MEMS	Microelectromechanical systems
MHD	Magnetohydrodynamics
MPA	Maritime patrol aircraft
MPF	Maritime prepositioning force
M&S	Modeling and simulation
MTCR	Missile Technology Control Regime

NATO	North Atlantic Treaty Organization
NIMA	National Image and Mapping Agency
NRO	National Reconnaissance Office
NSA	National Security Agency
NSFS	Naval surface fire support
NSS	Naval Simulation System
NTACMS	Navy Tactical Missile System
OMFTS	Operational Maneuver From the Sea
OOTW	Operations other than war
OSD	Office of the Secretary of Defense
PEBB	Power electronic building block
QOL	Quality of life
R&D	Research and development
RAMICS	Rapid Airborne Mine Clearance System
SABRE	Rocket-propelled line charge (mine warfare)
SAM	Surface-to-air missile
SAR	Synthetic aperture radar
SOF	Special operating forces
SQUID	Superconductor quantum interference device
SSBN	Nuclear-powered ballistic missile submarine
START	Strategic Arms Reduction Treaty
STOL	Short takeoff and landing
STOVL	Short takeoff and vertical landing
SZ	Surf zone
TBM	Tactical or theater ballistic missile
UAV	Unmanned aerial vehicle
UCAV	Uninhabited combat air vehicle
UGV	Unmanned ground vehicle
USAF	U.S. Air Force
UUV	Unmanned underwater vehicle
VLS	Vertical launch system
VSTOL	Vertical and short takeoff and landing